Lecture Notes in Economics and Mathematical Systems 594

Reinhard Hübner

Strategic Supply Chain Management in Process Industries

An Application to Specialty Chemicals Production Network Design

With 57 Figures and 22 Tables

Springer

Reinhard Hübner
McKinsey & Company, Inc.
Kurfürstendamm 185
10707 Berlin
Germany
Reinhard_Huebner@mckinsey.com

This book is the published version of the doctoral dissertation "Production network design in specialty chemicals" approved by the Faculty VIII - Economics and Management of the Technical University Berlin (D 83).

Library of Congress Control Number: 2007926109

ISSN 0075-8442

ISBN 978-3-540-72180-2 Springer Berlin Heidelberg New York

Springer is a part of Springer Science+Business Media

springer.com

Production: LE-TEX Jelonek, Schmidt & Vöckler GbR, Leipzig
Cover-design: WMX Design GmbH, Heidelberg

SPIN 12054275 88/3180YL - 5 4 3 2 1 0 Printed on acid-free paper

Preface

Working on operational performance improvement projects in chemical industry gave me the opportunity to experience first hand the challenges this industry is faced with due to changes of the competitive landscape and the shift of demand growth to the developing regions of the world. Addressing these challenges requires not only an operational but also a strategic response. The production (network) strategy is at the heart of the problem for many companies. Decision makers in industry are aware of the need to adapt their production networks but lack adequate methodological support to holistically re-design them. The core objectives of my endeavor into academic research were to develop a comprehensive approach towards specialty chemicals production network design and to demonstrate the insights the use of OR methods can provide in strategic planning problems.

Achieving these objectives would not have been possible without the support of a large number of people. First and foremost, I would like to thank my academic mentor, Professor Dr. Hans-Otto Günther, Technical University of Berlin. He evoked my interest in production and operations management when I was a student at his department and wholeheartedly supported my dissertation project. My work benefited significantly both from his personal feedback and the frequent discussions within the department. Throughout the completion of the dissertation, Professor Dr. Martin Grunow, now at the Technical University of Denmark in Copenhagen, has always been an excellent sparring partner. Markus Meiler and Jenny Golz gave me very valuable tips for programming the optimization model and Boris Otte implemented the scenario and sensitivity analysis features as part of a student project.

The close cooperation with a global specialty chemicals company that whishes to remain anonymous made it possible to link academic research with application-oriented considerations. I would like to express my gratitude towards all employees of that company who enthusiastically supported my work.

Last but not least, I would like to thank my fiancé Karin Hellner for bearing with me throughout all phases of this journey and my brother Rudolf for proofreading the manuscript.

Reinhard Hübner, Berlin, March 2007

Contents

List of Abbreviations

ABC	Activity Based Costing
AHP	Analytic Hierarchy Process
ANP	Analytic Network Process
APS	Advanced Planning Systems
ATP	Available To Promise
BOM	Bill Of Materials
CAPM	Capital Asset Pricing Model
CTP	Capable To Promise
DEA	Data Envelopment Analysis
DMU	Decision Making Unit
EDR	Expected Downside Risk
ERP	Enterprise Resource Planning
FTE	Full-Time Equivalent
GATT	General Agreement on Tariffs and Trade
GDP	Gross Domestic Product
KPI	Key Performance Indicator
MADA	Multiple Attribute Decision Analysis
MAUT	Multiple Attribute Utility Theory
MAVT	Multiple Attribute Value Theory
MCDA	Multiple Criteria Decision Analysis
MFN	Most Favored Nation
MILP	Mixed-Integer Linear Programming
MINLP	Mixed-Integer Non-Linear Programming
MIP	Mixed-Integer Programming
MODA	Multiple Objective Decision Analysis

NAFTA	North American Free Trade Area
NPV	Net Present Value
OECD	Organization for Economic and Commercial Development
OR	Operations Research
SMART	Simple Multiple Attribute Rating Technique
STN	State-Task-Network
U.S.	United States
UFLP	Uncapacitated Facility Location Problem
UPM	Upper Partial Mean
VCI	Verband der Chemischen Industrie e.V.
WACC	Weighted Average Cost of Capital
WTO	World Trade Organization

1 Introduction

1.1 Motivation and Objectives

Globally, chemical industry realized revenues of 1,776 billion Euros in 2004. The European Union is the world's largest producer of chemical products with a 33% share of global production and with revenues of approximately 140 billion Euros. Germany, where chemical industry represents approximately 10% of total industrial output, holds the third rank surpassed only by the United States and Japan. Additionally, both the world's largest chemical company BASF and the world's largest specialty chemicals company Degussa are headquartered in Germany. While worldwide chemical production is concentrated to a few countries (the top 10 countries represent more than 70% of global production), the industry is nevertheless truly global from an operations perspective. On the one hand, international trade represents more than 40% of global revenues. On the other hand, major chemical companies typically operate numerous production sites in all major economic regions of the world. With 124 billion Euros in 2004, the revenues generated by the international subsidiaries of German chemical companies almost equal the revenues generated from domestic operations.[1]

Historically, as was the case in many other industries, foreign production sites were primarily established to access local markets (cf. Dasu and de la Torre 1997, p. 313). This strategy is typically associated with a duplication of manufacturing operations (cf. Shi and Gregory 1998, p. 205). However, trade barriers such as high tariffs made local production assets a prerequisite for accessing foreign markets. Additionally, today's industry has been shaped by numerous mergers and acquisitions (cf. Lesney 2004; Mullin 2004; Storck 2004). The comprehensive supply network integration effort normally required after a merger or acquisition (cf. Goetschalckx and Fleischmann 2005, p. 117) in many cases did not take place in chemi-

---

[1] Data for this paragraph was obtained from several publications of the industry organization of the German chemical industry (Verband der Chemischen Industrie e.V. or VCI); cf. VCI 2006, VCI 2005.

cal industry. As a consequence of these factors, the supply networks found today often lack a coherent design strategy.

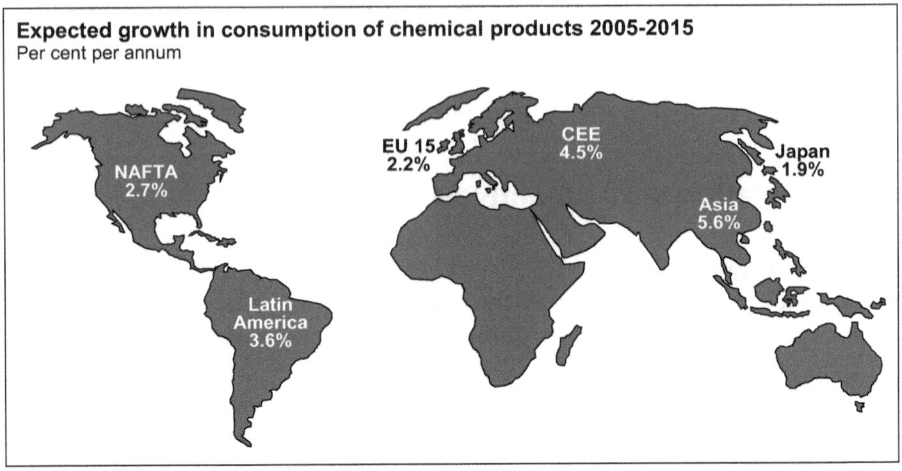

Fig. 1. Expected growth of chemicals consumption (source: VCI 2004, p. 2)

The competitive situation of chemical industry has been driven by cost pressures for several decades (cf. Riggert 1992). The *price-cost squeeze* caused by rising raw material prices that cannot be fully passed on to customers, already observed since the 1980ies (cf. Bartels et al. 2006, p. 96), has recently become even more pronounced. Data from Germany's industry organization VCI illustrates this point: in Germany between 2000 and 2005 the price index of chemical products grew by 5.8% while the raw material price index, primarily driven by rising oil and gas prices, grew by 72.5% (cf. VCI 2006, pp. 24-31). At the same time competition has become truly global. The removal of trade barriers has contributed to the emergence of contenders from low cost countries especially in Asia (cf. Nickel 2006, pp. 244-245). Hence, the need to achieve a competitive cost position has become more pronounced. Additionally, matters are complicated by the fact that markets in the major industrialized countries are relatively stagnant and the strongest growth is happening in countries such as India and China (cf. Fig. 1). As Bartels et al. (2006, p. 100) point out, this is not only because of differentials in GDP growth but also because many customer industries such as electronics, textiles and automotive industry are migrating operations to Asia. Of the 120 chemical plants with investments exceeding US$ 1 billion currently under construction worldwide, 50 are located in China (cf. The National Academies 2006, p. 9). Summing up the major supply chain challenges process industries are faced with, Shah (2005, p. 1225) expects even more dynamic and competitive markets, shorter product life cycles and the need to deliver specialty products at commodity prices via mass customization.

Companies in chemical industry employ different levers to improve their competitive position. Common approaches include overhead cost reduction efforts, step-change productivity improvement programs and the implementation of continuous improvement efforts across all functions. The need to manage global supply networks taking an integrated perspective, already postulated to be a major challenge for manufacturing companies by McGrath and Hoole (1992, pp. 94-95), is also taking hold in chemical industry (cf. Nickel 2006, p. 247; Hartmann et al. 2001). As Ferdows (1997a, p. 109) points out, doing so can in itself be the source of competitive advantages. This is also confirmed by empirical research showing that adapting the supply network to changes in the competitive environment is a critical success factor (cf. Lee 2004). Yet, the improvement potential from comprehensively re-designing entire supply networks has so far received insufficient attention (cf. Vallerien and Wittemann 2002, p. 17), despite the fact that global supply networks offer the opportunity to actively exploit comparative advantages from regional differences in capabilities, factor costs, market potentials, etc. (cf. Cohen and Mallik 1997, p. 194; Porter 1990, pp. 577-616). Only recently have companies in chemical industry begun to restructure their supply networks e.g., via plant closures to eliminate overcapacities or transfers of entire product lines to low-cost countries such as China (cf. N.N. 2004; Pollak 2002).

Advanced Planning System (APS) vendors have supported industry in its efficiency improvement efforts by providing tools and methods to improve planning processes and facilitate an integrated management of entire supply networks. Company-internal efforts have been supplemented by increased cooperation with suppliers and customers and coordination across multiple value chains (cf. Cohen and Huchzermeier 1999, p. 671), made possible by state-of-the art software and internet technology. However, the design of the supply networks has often been beyond the scope of these improvement activities. As Daskin et al. (2005, pp. 39-40) point out, facility location and capacity selection decisions, due to their long-term consequences and the great amount of interdependencies, are the most difficult ones within supply network design. A lack of customized analytical tools to support these decision processes may be one important reason for the reluctance to tackle supply network redesign in chemical industry. The network design modules provided by APS vendors have so far not been capable of providing generic network design models that can be sufficiently customized to model complex production networks from industry and

Günther and Tempelmeier (2005, p. 332) even question the usefulness of trying to do so.

Supply network design has been in the focus of the academic operations research community for many years. However, as Vos and Akkermans (1996, p. 58) point out, most publications solely focus on the formulation of optimization models and ignore the integration of their models into management processes and decision support tools. Additionally, the majority of the models proposed is of a general nature and hence does not accommodate industry-specific requirements. From the 77 optimization models reviewed in Chapter 3.2.1 only 12 contain features specific to an application industry. While the sensitive nature of network design issues might lead to a reluctance to report application experiences (cf. Cohen and Mallik 1997, p. 205), Eiselt (1992, p. 5) concedes that a theory-practice gap exists in supply network design research. This gap might also be responsible for the fact that mathematical programming techniques are often mistakenly presumed to be too complex to be applied in industry (cf. Vidal and Goetschalckx 2000, p. 101).

The objective of this work is to contribute towards closing this gap. To this end, quantitative and qualitative tools required to design and especially re-design production networks in specialty chemicals industry are developed and integrated into a comprehensive planning process. Cornerstone of this work is a Mixed-Integer Linear Programming model to support production network design analysis and optimization. In developing the optimization model, the focus is not on creating new Operations Research methods but on capturing the economic and technical aspects of production network design problems from industry. Insofar an important contribution of this work will be to demonstrate how Operations Research methods can be applied to support strategic planning processes in industry and illustrate the insights that can be gained from doing so. To achieve these goals and ascertain that major issues practitioners from industry are faced with are resolved, this research project was conducted in cooperation with a European specialty chemicals company which operates a production network of more than 50 sites spread across all continents.

1.2 Approach and Dissertation Outline

The dissertation consists of 5 chapters in addition to this introduction. Chapter 2 lays the foundation by establishing the role of production network design within supply chain management. To this end key terms are defined, the role of Advanced Planning Systems in production network design is discussed and core concepts from manufacturing strategy research related to production network design are presented. Subsequently, the links between production network design and the research on industrial location science are established and the characteristics of chemical industry in general and the peculiarities of the specialty chemicals segment are introduced. Finally, an integrated planning and controlling process for production network design in specialty chemicals industry is proposed and the analyses and decision support tools required in each phase are defined.

Chapter 3 deals with the global production network optimization phase based on employing Operations Research methods. First, a brief introduction to the general literature on facility location is provided and the literature on production network design is reviewed. Based on a comprehensive discussion of modeling variants from literature, modeling approaches tailored to the peculiarities of specialty chemicals industry are proposed for all critical elements of the respective production networks. Absorbing the merits of this discussion, a Mixed-Integer Linear Programming model is developed. Additionally, extensions to allow for the applicability of the model to a broader range of production systems than those considered in the course of the research project are provided. The model formulations especially focus on capturing the economic questions underlying the network design problem. Results from numerical tests are given to demonstrate that commercial optimization software is capable of solving the proposed model for problem instances of realistic size.

Chapter 4 covers the site selection and site controlling phase. Consequently, it deals with the assessment of individual production sites based on primarily qualitative criteria. Alternative Multiple Attribute Decision Analysis methods are reviewed and a decision support model employing the Analytic Hierarchy Process, which can be used both for site selection problems and as a controlling tool to perform site portfolio rankings of entire production networks, is proposed. Experiences from application in industry are reported.

An application case study of the production network optimization model is reported in Chapter 5. In this context the integration of the optimization model into a planning tool to support interactive explorations of the solution space is demonstrated and guidance on how to develop the data required for quantitative strategic network design analyses is provided. Additionally, important analyses that can be performed using the proposed optimization model are introduced and improvement potentials identified in the course of a pilot application in industry are explained.

To conclude the dissertation, Chapter 6 summarizes the key findings of this work and provides directions for future research.

2 Production Network Design and Specialty Chemicals

2.1 Supply Chain Management and Production Network Design

2.1.1 Supply Chains and Production Networks

Many different definitions of the term *supply chain* exist in literature (cf. Ganeshan et al. 1999, p. 842). Christopher (2005, p. 17) defines the supply chain as a "...network of organizations that are involved, through upstream and downstream linkages, in the different processes and activities that produce value in the form of products and services in the hands of the ultimate consumer". Typically, a supply chain consists of suppliers, production sites, storage facilities, distribution facilities and customers linked by material, information and financial flows. As shown in Figure 2, a supply chain can be spread across several facilities located in different countries that might belong to different companies. At the same time, depending on the product portfolio, a company is usually part of numerous supply chains (cf. Lambert and Cooper 2000, p. 69).

The resulting network of interlinked facilities/organizations is also referred to as *supply network* (cf. Günther 2005, p. 5). Its overall complexity is largely determined by the number of *echelons* (inventory carrying facilities) included (cf. Tsiakis et al. 2001, p. 3585), but global spread may also add significant additional complexity. Within the supply network one can distinguish between the *production network* and the *distribution network*. While the production network consists of all production facilities and the inventory facilities required for their operation, the distribution network consists of all inventory and distribution facilities required to deliver products to final customers.

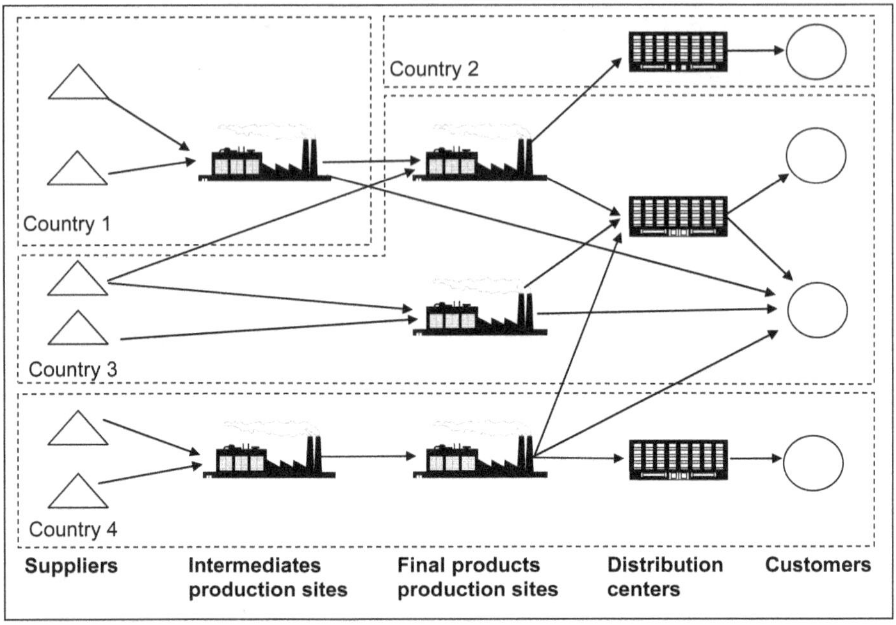

Fig. 2. Global supply chain network

Supply chains technically range from the extraction of raw materials to the final customer and are usually spread across several companies. Narrowly defined, the supply chain is limited to elements operated by an individual company (*intra-organizational supply chain*), whereas a broad definition also includes elements operated by other parties, also referred to as *inter-organizational supply chain* (cf. Stadtler 2005, pp. 9-10; Shah 2005, p. 1226). The different definitions also reflect the fact that, as explained by Rudberg and Olhager (2003), with operations management and logistics management two major research tracks merged different perspectives into what is nowadays referred to as supply chain management. Operations management, taking an intra-company perspective, originally focused on the manufacturing nodes of the network while logistics management focused on the material and information flows between the nodes of the network and, including the inter-company perspective, the flows between the network and suppliers/customers.

Following the rationale of Shi and Gregory (1998, p. 199) that a company should first optimize the elements of a supply chain under its own control, issues related to the inter-organizational integration and coordination of supply chains will generally not be covered in this work. For an overview of specific inter-organizational supply chain management tasks the reader can refer to textbooks such as Chopra and Meindl (2004) or Simchi-Levi et al. (2003). A specific discussion of inter-organizational aspects of supply chain management and further references can for example be found in Kilger and Reuter (2005), Kuhn and Hellingrath (2002) and Chen (2003) with the latter focusing on the benefits of information sharing between supply chain partners.

2.1.2 Production Network Design

Supply chain management has to address diverse issues ranging from facility location to detailed production and procurement decisions (cf. Fleischmann et al. 2005a, pp. 86-92). To reduce the complexity of the planning process, planning activities can be hierarchically decomposed based on their time horizon and their importance for the company. While some authors distinguish only between strategic and operational planning, the framework most commonly employed - originally proposed by Anthony (1965) - includes tactical planning as an intermediate level. In the context of supply chain management, the different planning levels can be defined as follows:[2]

- *Strategic* planning focuses on creating and sustaining the conditions required for the successful long term development of a company. The time horizon usually covers a period of three to ten years. Decisions are of great importance for the company and typically include among others the product and services portfolio, configuration of production and distribution networks and investments into new production technologies.
- *Tactical* planning lays out a step by step approach for the implementation of the strategic objectives with a time horizon of 1 to 3 years. Typical decisions include the launch or discontinuation of specific products, production capacity adjustments and product transfers within the existing production network.
- *Operational* planning ensures the optimum utilization of existing assets and efficient execution of the decisions taken in strategic and tactical planning. The time horizon covered is up to one year with daily or weekly intervals. Typical decisions include detailed production scheduling and distribution scheduling.

[2] cf. Günther and Tempelmeier (2005), p. 27; Chopra and Meindl (2004), pp. 7-8; Simchi-Levi et al. (2003), p. 15; Miller (2002), pp. 2-6; Schmidt and Wilhelm (2000); Zäpfel (2000), pp. 1-16.

While there are strong interdependencies between strategic, tactical and operational planning, in practice planning processes generally take place in a hierarchical fashion with strategic planning forming the basis of the different operational plans (cf. Hahn 1992, col. 1988-1991; Hax and Meal 1975, p. 54). In the field of supply chain planning first models combining strategic and operational planning have been proposed, e.g., Kallrath (2002) or Sabri and Beamon (2000). However, as other authors such as Fleischmann and Meyr (2003, pp. 475-477), Miller (2002, pp. 7-8) and Bitran and Tirupati (1993, p. 525) argue, an integrated approach is not necessarily desirable because strategic, tactical and operational planning have to deal with different degrees of uncertainty, different planning horizons and corresponding planning frequencies, different aggregation levels and ultimately decisions are of different degrees of importance. Taking the latter position, this work focuses on strategic supply chain planning which is also referred to as *supply network design*. Elements of tactical planning will be covered if required, e.g., in the context of reallocation of production volumes within an existing network to react to exchange rate fluctuations.

Production network and *distribution network* design are closely interlinked elements of supply network design. In literature, both combined production/distribution models and separate models for either production or distribution network design are proposed. For example, Ambrosino and Scutellà (2005), Simchi-Levi et al. (2003, pp. 23-42) and Muriel and Simchi-Levi (2003) focus on distribution network design. On the other hand, authors such as Nickel et al. (2005), Wouda et al. (2002) and Canel and Khumawala (1997) focus on production network design while Melo et al. (2005), Goetschalckx et al. (2002) and Arntzen et al. (1995) propose integrated models. Whether an integrated approach is required or a focus on the production network is sufficient primarily depends on the relative importance of transportation costs. A good indicator for an initial assessment is the value density of the products (monetary units per weight/volume unit). For example, Camm et al. (1997, p. 132) justify their decomposition approach by pointing out that in process industries production and material costs often dominate distribution costs. This is in line with results from the pilot application reported in Chapter 5 where distribution costs were in the range of 2-4% of total costs and thus well below the level of most other cost factors. The majority of these were captured by modeling the transport processes from producing site to destination country without explicitly considering further distribution echelons involved.

Besides the issue of cost relevance, interdependencies between production and distribution networks are often limited for companies already operating global networks. Distribution facilities usually serve major markets and their location and capacity are fairly independent of individual plant location decisions. Consequently, this work focuses on production network design. An overview of recent publications on distribution network design is for example provided by Klose and Drexl (2005).

Major decisions to be made when designing a production network are:[3]

- Whether to operate only one site or split production across several sites,
- The definition of the production network's geographical footprint (e.g., only in one country, only within one economic area or global operations),
- The underlying design principle of how to split production across a multi-plant net-work and the integration of individual sites into the overall production network,
- And the number, location, capacity and technology of sites including allocation of products/product variants and markets to individual sites (partly determined by the chosen network design principles).

While the first three points are important aspects of production network design, they are in the majority of cases predetermined by the fact that companies already operate a global production network. Even medium-sized companies in many industries operate production sites outside their home country to have access to international markets or benefit from comparative advantages. Consequently, this work gives a brief introduction to network design principles (third bullet point) while focusing on physical network design. This is at the same time clearly the most complex part of production network design as number, location, capacity and technology decisions within a network are highly interdependent and thus require simultaneous planning (cf. Chopra and Meindl 2004, p. 99; Dasci and Verter 2001, p. 963). As pointed out by Verter and Dincer (1995, p. 265) due to location-specific differences in the availability and cost of production factors these interdependencies are even more pronounced in an international context.

Also, it should be noted that in practice, as stressed for example by Harrison (2003, p. 5), a redesign of existing production networks, initiated in the course of mergers and acquisitions, strategy changes or capacity adjustments, is much more common than the design of a new production network in a greenfield approach. Therefore, this work specifically incorporates issues arising from redesign of existing networks such as restruc-

---

[3] cf. Shah (2005), p. 1226; Chopra and Meindl (2004), p. 99; Simchi-Levi et al. (2003), p. 15; Tsiakis et al. (2001), pp. 3585-3586; Neumann et al. (2002), pp. 254-256; Goetschalckx (2000), p. 79; Götze (1995), pp. 50-51; Verter and Dincer (1995), pp. 264-265.

turing costs and limitations on the degree of freedom imposed by the existing network. As the design of new production networks will also be dealt with, in the remainder, unless explicitly noted, design and redesign will be used synonymously.

2.1.3 Production Network Design and Advanced Planning Systems

According to Fleischmann and Meyr (2003, p. 457) adequate planning systems for supply chain management require two major elements:

- An *integral planning* of a company's entire supply chain including at least suppliers and customers while taking into account the interdependencies between the various activities
- A *true optimization* of decisions based on exact or heuristic optimization algorithms

The material requirements planning incorporated into commonly used Enterprise Resource Planning (ERP) software does not contain this type of planning functions (cf. Tempelmeier 1999, p. 69; Drexl et al. 1994). To address this deficit, software developers introduced so-called Advanced Planning Systems (APS) that incorporate these two elements based on a hierarchical planning concept. The different APS, though developed independently by several software companies, have a common underlying structure (cf. Meyr et al. 2005). Figure 3 displays the software modules usually found in an APS. Even though there are strong links between the modules, companies using APS can decide which modules to use depending on the needs of their individual supply chains. Additionally, APS vendors developed a range of industry-specific modules. An explanation of the different modules and application examples can be found in Günther (2005, pp. 12-37).

As Günther (2005, p. 15) points out, the strategic network design module of commercial APS, while attempting to cover the full range of supply network design tasks, is in practice probably the least utilized module of APS. In line with this finding, Hurtmanns and Packowski (1999) do not even discuss this module in their paper on the deployment of APS in chemical industry. According to Grunow et al. (2006, p. 1) the lack of deployments in industry has even led some major vendors to cease promoting the network design modules of their APS systems. This development can amongst other reasons be attributed to the fact that network design issues arise only infrequently, that problems are too particular for "generic" APS and that more specific models can be developed based on commercial optimization software (cf. Fleischmann et al. 2006, p. 8). Additionally, the data integration with the ERP system, a key advantage of using APS instead of standalone tools, is rather low for strategic network design (cf. Goetschalckx and Fleischmann 2005, p. 133).

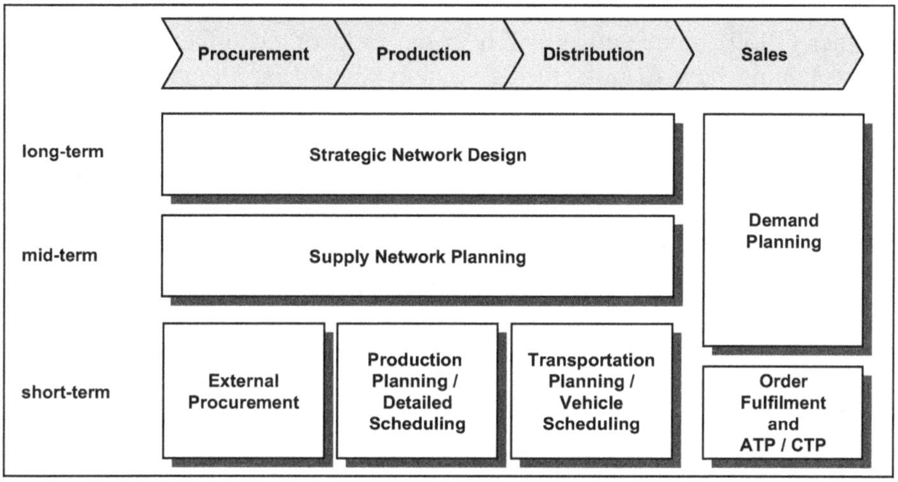

Fig. 3. Software modules of APS[4]

An application employing the strategic network optimization module SNO from Oracle (formerly J.D. Edwards) at BMW AG as reported by Henrich (2002) illustrates these limitations. When BMW wanted to extend the model to include investment decisions it turned out that the problem became both too particular and too complex to be modeled using the commercial APS tool and BMW reverted to creating the model using optimization software supplied by ILOG (cf. Fleischmann et al. 2006, p. 8). Therefore, the use of commercial APS for production network design will not be further pursued. The reader interested in details on APS solutions in process industry is referred to Günther and van Beek (2003).

4 Source: Günther (2005), p. 10. The structures of particular APS are discussed in Fleischmann and Meyr (2003), pp. 509-516.

2.1.4 Generic Production Network Design Strategies

Manufacturing Strategy and Production Network Design

Production network design is a central element of manufacturing strategy. Skinner (1969) pioneered research in the field of manufacturing strategy by explaining how manufacturing strategy should be aligned with corporate strategy. To remain within the scope of this work, only manufacturing strategy research findings specifically dealing with production network design will be introduced below. Further references on manufacturing strategy in general are provided for example in Dangayach and Deshmukh (2001) and a case study describing the development of a manufacturing strategy aligned with overall business strategy can be found in Beckman et al. (1990).

Research on production network design strategy can also be traced back to Skinner (1974). He developed the concept of the *focused factory* based on the insight that a factory cannot perform well on all types of manufacturing performance metrics simultaneously (according to Spring and Boaden (1997, p. 758) the relevant metrics are cost, quality, delivery dependability, delivery speed and flexibility). Instead, factories have to be focused based on the competitive priority defined by corporate strategy. As Skinner suggests, focus can be achieved either by operating separate facilities for each type of competitive priority or by implementation of the "plant within a plant" concept whereby a large facility is divided into independent units focused on their respective competitive priorities. A typical approach towards increasing focus is for example to reduce the product variety produced at each facility. Stalk (1988, pp. 42-43) gives examples of cost savings that can be achieved with this approach.

The result of aligning production network design with business strategy is a distinct production network design strategy for each business. Core elements of network strategy, namely the segmentation principle, the strategic role of a plant within the network and flexibility considerations are described below. Broken down to the plant level, the result is a *plant charter* that describes the role of the respective plant within the overall production network (cf. Hayes and Wheelwright 1984, p. 100).

For existing production networks, empirical research conducted by Vokurka and Flores (2002) shows that the link between network strategy and competitive priorities is often missing. McGrath and Hoole (1992, p. 100) even concede that corporate management at some companies simply lacks the power to align regional operations around a consistent manufacturing strategy. One reason for this observation might be that production networks in many cases were not developed from scratch but grew historically with many sites being added in the context of merger and acquisition activities (cf. Küpper 1982, p. 443). Consequently, significant improvement potential can be expected from an optimization of existing production networks.

General Network Design Principles

If a company decides to operate more than one site, it has to decide on how to distribute activities across its sites. Ihde (2001, pp. 85-87) describes the basic options available. One option is to split volumes so that all sites perform all activities. This option basically duplicates activities at each new site. A second option is to divide activities across several sites by function, product or production process. In this case, each site specializes on a specific segment of the overall activities spectrum. Finally, the two options can also be combined leading to what Ihde calls a diversified site network.

Considering only production network design, Schmenner (1979) builds on the focused factory concept to develop four distinct multi-plant strategies. While he does not consider an international environment, the generic strategies developed for domestic networks are also applied to international production networks (cf. Kouvelis et al. 2004, p. 127). Based on a product/market or process focus Schmenner defines four plant types:

- *Product plants* serve the company's entire market for the products they produce specializing on the competitive priorities associated with their product portfolio.
- *Market area plants* produce a majority of the company's products for distribution to their regional market.
- *Process plants* focus on certain process steps usually with some plants providing components for other plants. They focus on the specific manufacturing requirements of certain components.
- *General purpose plants* are designed for flexible assignment of products, markets and process segments without a specific focus.

Strategies can also be combined, e.g., by establishing product plants in each of the major economic regions. Hayes and Wheelwright (1984, p. 91) and Dornier et al. (1998, pp. 259-262) list some of the advantages and disadvantages associated with Schmenner's network strategies. Kulkarni et al. (2004) show that a process plant strategy can have risk pooling advantages even in the absence of economies of scale. For the United States, Schmenner (1982a) found product and to a lesser extent market area plants to be by far the most common strategies. Similarly, international plants were typically added as market area plants leading to a replication of activities (cf. Cohen and Kleindorfer 1993, p. 12). Comparing their findings 20 years later with Schmenner's data, Vokurka and Flores (2002) observe a strong trend away from market area plants towards integrated production networks which can be based on a product or a process focus. This trend was also observed by Flaherty (1986). McGrath and Hoole (1992, p. 95) even state that this integration is a must to survive in global competition.

Strategic Plant Roles

Ferdows (1989) uses the primary reason for establishing a plant (cheap production factors, use of local technological resources or proximity to markets) and the extent of technical activities taking place at the plant to distinguish between six strategic plant roles in international production networks:

- *Off-shore factories* utilize local factor cost advantages to supply components or final products to the home plant.
- *Outpost factories'* primary role is to collect information on advanced suppliers, competitors, research laboratories or customers. Ferdows considers them to be a theoretical option.
- *Source factories* are established primarily to benefit from cheap production factors. In contrast to off-shore factories they additionally become a focal point for certain production processes, components or products.
- *Server factories* are established to serve specific national or regional markets.
- *Contributor factories* combine source and server factory principles. They primarily serve specific national or regional markets but also become focal points for certain production processes, components or products.
- *Lead factories* are located in regions with local technological resources to build strategic manufacturing capabilities. They usually are the sole or major production resource for certain products and components in the company's production network.

In addition to its primary role a factory can also have a secondary strategic role. This can either be another one of the roles described above but could for example also be that it provides operational hedging against currency risks. In order to establish manufacturing as a source of competitive advantage, Ferdows (1997b) argues that companies should strategically develop a factory's role within the production network. While some factories might keep their original role for a long period of time, generally Ferdows, taking a "resource-based view" perspective on site planning, assumes that upgrading the strategic role of a factory offers competitive advantages. Figure 4 shows the possible paths to higher strategic roles as described by Ferdows. Further implications of the resource based view on site planning are discussed in Götz and Mikus (2002).

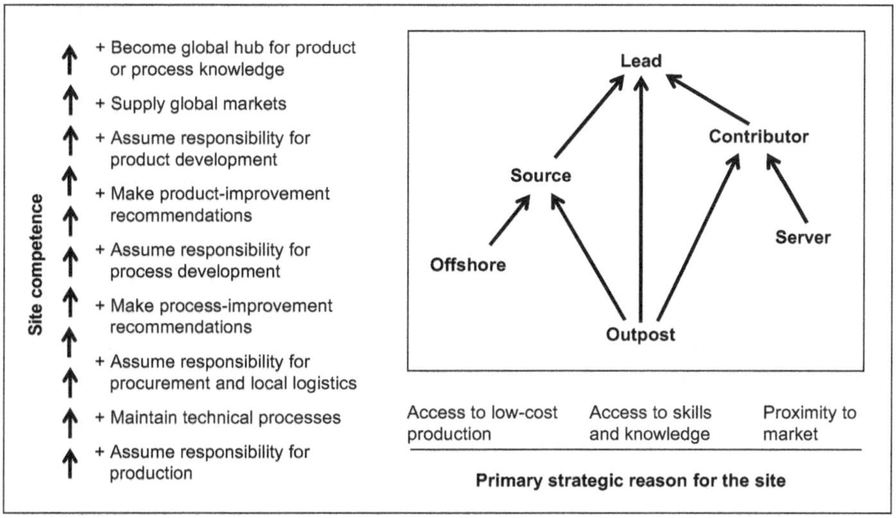

Fig. 4. Paths to higher strategic roles (cf. Ferdows 1997b, p. 79)

In an empirical study Vereecke and Van Dierdonck (2002) tested Ferdow's model and found it to be valid with two exceptions: it appears to be too limited in the criteria for adding plants to an existing network and lead factories were also added based on market proximity. In another study Maritan et al. (2004) used autonomy over planning, production and control decisions to validate Ferdow's model but found only weak correlations with planning decisions showing the strongest correlation.

As De Meyer and Vereecke (2001) point out, the frameworks proposed by Schmenner and Ferdows approach production network design strategy from different angles. While they provide complementary insights, they are not mutually exclusive. The question of which network design strategy is most appropriate for a business has to be answered based on the competitive priorities and the characteristics of the product portfolio and production processes of the respective supply chain. Furthermore, it should be noted that network design can also be analyzed taking an industry perspective.

Flexibility and Production Network Design

Flexibility considerations have become increasingly important in the context of designing production systems in all industries (cf. Bertrand 2003, p. 133; Upton 1995, p. 74) and it has been argued for some time that flexibility can in itself be the source of competitive advantages (cf. Beckman 1990). Various definitions exist in literature (cf. Chambers 1992, pp. 287-291), but the basic elements constituting flexibility as defined by Skinner (1985, pp. 43-44) are process, product and volume flexibility. *Product flexibility* refers to the ability to accommodate changes in the product mix and product characteristics while *volume flexibility* indicates the ability to accommodate volume changes due to seasonality of demand, product lifecycles, etc. Finally, *process flexibility* refers to the ability to adapt to changes in processing requirements. A comprehensive discussion of different types of flexibility, their interrelationships and how to measure and achieve them can be found in Sethi and Sethi (1990).

In chemical industry, due to the nature of chemical production processes, product and process flexibility are much more limited by technological constraints than in other industries. Hence, a discussion of flexibility choices arising at the equipment/plant level will not be pursued here. Instead, selected findings from research on flexibility issues arising in supply network design, with a focus on volume flexibility, are introduced below. Bertrand (2003) provides a general overview of the subject of flexibility in supply chain design where further references to literature and decision models to deal with resulting trade-offs can be found.

Jordan and Graves (1995) analyze volume flexibility that can be achieved via product-plant assignment choices in a multi-plant, multi-product production network when faced with uncertain demand. Based on a 10 plants/10 products example they demonstrate that, if correctly designed, a network with partial flexibility can yield almost the same volume flexibility benefits as a totally flexible network where all plants are able to produce all products. Their recommendation is that products should be allocated to plants in a "chain pattern" with the complete network ideally creating a single chain instead of several shorter chains (cf. Fig. 5). For more complex networks their recommendation is to equalize the number of plants a product is directly connected to and the number of products to which each plant is directly connected and create a circuit that encompasses as many plants and products as possible.

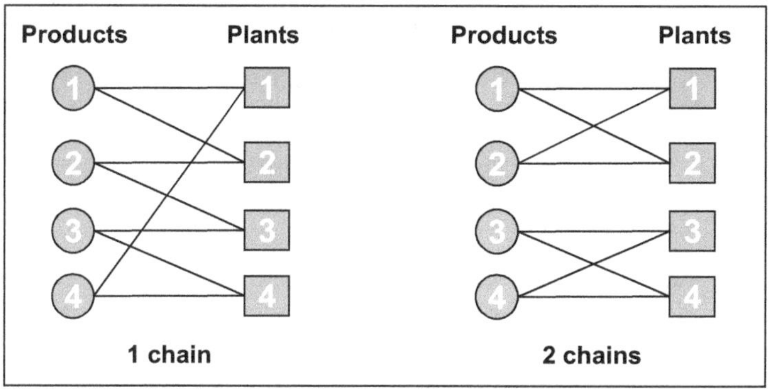

Fig. 5. Chain configuration of production networks[5]

An international production network offers an additional set of flexibility opportunities because it allows companies to respond to shifts in comparative advantages caused by events such as changes in government policies, decisions of competitors or exchange rate fluctuations (cf. Porter 1986, p. 21). Kogut (1985) identifies six sources of competitive advantage offered by an international network: arbitrage opportunities from production shifting, tax minimization, financial markets and information arbitrage, leverage opportunities from global coordination and leverage against political risks. The most well-researched aspect in this area is probably the effect of exchange rate fluctuations where the use of real options theory is often proposed to assess the financial value derived from operational flexibility. The issue of exchange rate risks is discussed in greater detail in Chapter 3.3.5.

2.2 Production Network Design and Industrial Location Science

2.2.1 Introduction to Industrial Location Science

As discussed above, production network design includes several interdependent decisions such as number, location, capacity and technology of plants. While all of these decisions have to be made simultaneously to create an optimized production network, the location decision is at the heart

[5] Source: Jordan and Graves (1995), p. 582.

of production network design. Location science, also referred to as facility location research or location analysis, investigates the spatial location of objects and usually either tries to explain observed locations or proposes processes and methods to determine locations that have certain properties. The types of facilities to be located include among others emergency services, communication systems, retail outlets and of course industrial production facilities.

Location science has received considerable attention in various academic disciplines including geography, economics, business administration, operations research and engineering. Two compilations of literature on the subject may serve as an illustration of the amount of research published: a bibliography on location and layout planning compiled by Domschke and Drexl (1985) already contains 1,800 entries and an online reference list on location science compiled by Trevor Hale contains more than 3,400 entries[6]. Additionally, practically every textbook on operations management or supply chain management contains a section on facility location. Obviously, a comprehensive review of the literature on location science is beyond the scope of this work. Therefore, this chapter is limited to a brief introduction to industrial location science and the discussion of a few seminal publications. Further references are included in the respective chapters of this work to avoid cross-references.

Meyer-Lindemann (1951) proposed the first segmentation of industrial facility location research in German literature. Based on the research objective, he defined four segments: *Location selection* analyzes the determinants of location decisions, *location impact* analyzes the effect a given location decision has on its environment, *location development* analyzes the historical development of site structures and *location policy* seeks to create economic policy options to influence location decisions of companies.

Meyer-Lindemann's initial classification has been refined and extended by various authors. Goette (1994, p. 50) and Autschbach (1997, p. 126) added *location planning* (dealing with the planning process of location selection) and *international location* (dealing with specific aspects of international location decisions). Figure 6 shows a simplified version of the classification Kaiser (1979, pp. 18-23) developed based on Behrens (1960). Kaiser distinguishes between economic location science and business location science. While economic location science analyses the distribution of industrial facilities across geographical space (descriptive) or develops policy options to influence this distribution (prescriptive), business location science explains location decisions of individual companies (descriptive) or develops models to support individual location decisions (pre-

[6] Cf. http://gator.dt.uh.edu/~halet/ (as of 2007-03-04).

scriptive).[7] Empirical location science is included as a separate field of research because both economic and business location science draw on the results of empirical analyses.

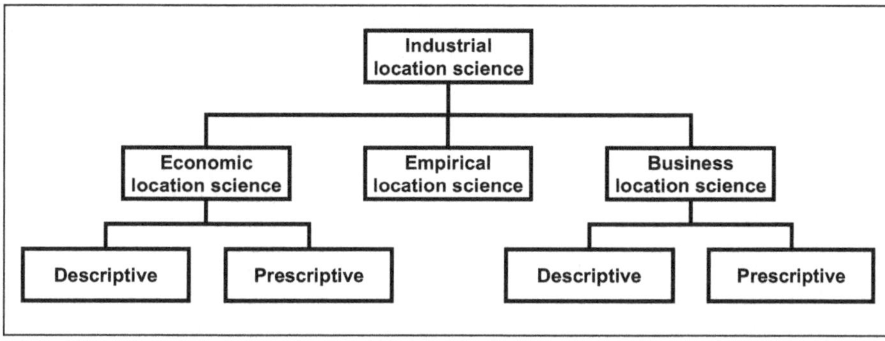

Fig. 6. Segmentation of industrial location science

Findings from both empirical location science and prescriptive business location science are relevant in the context of production network design. Prescriptive business location science provides factors relevant in plant location decisions, quantitative and qualitative methods to evaluate alternative sites and systematic facility location planning processes. Empirical research both on location selection and location planning provides insight into how industry actually approaches location decisions. Sometimes, layout planning is also included in the definition of business location science and some publications (e.g., Francis et al. 1992) cover both location and layout planning. Here, research on economic location science, descriptive business location science and layout planning will generally not be considered.

2.2.2 Major Findings from Industrial Location Science

Both in German and English literature on industrial facility location, Alfred Weber's publication "On the location of industries" ("Über den Standort der Industrien", Weber 1909) is commonly cited as the origin of industrial location science (cf. Eiselt and Laporte 1995, p. 151; Brandeau and Chiu 1989, p. 645). Weber introduces the concept of *location factors* which he defines as "factors constituting a precisely defined advantage that

[7] For the distinction between descriptive and prescriptive location science cf. also Krarup and Pruzan (1990), p. 2.

is realized if an economic activity takes place at a certain location or more generally at locations of a certain kind" (cf. Weber 1909, p. 16; translated by the author). The systematic of location factors he developed (cf. Weber 1909, pp. 18-22) distinguishes between general factors relevant for all types of industries and specific factors that are relevant only for certain industries. Based on the spatial effect of the factors, a further distinction between regional factors causing companies to locate at a specific location, agglomeration factors causing a concentration of industry in certain regions and factors leading to a decentralization of industry is drawn. Finally, he groups location factors into natural/technical factors and societal/cultural factors.

Weber reduces the relevant decision factors to transportation costs, labor costs and the effect of agglomeration factors (cf. Weber 1909, pp. 22-35). To analyze transportation costs, he categorizes materials into three different groups: materials that become part of the finished product with their full weight, materials that become part of the finished product with a share of their weight or not at all (e.g., coal in steel production) and ubiquities that are available everywhere and thus not relevant for location decisions (e.g., air). In a first step, Weber identifies the least transportation cost location. Labor costs and agglomeration factors are included in a second step by considering the compensation process between savings from labor cost/agglomeration advantages and additional transportation costs. For identifying the least transportation cost location in the case of two raw materials and one demand location, Weber suggests to use a mechanical device called a Varignon frame. It attaches weights representing the transportation quantities to three connected strands forming the location triangle. The connection of the strands is moved to the location with the lowest transportation costs by force of the weights. In doing so, Weber laid the foundation for the field of research focusing on the use of optimization models to support location decisions.

While the underlying mathematical optimization problem, also referred to as *Steiner-Weber-problem* or *minisum problem*, is one of the classical models discussed in operations research literature on facility location (cf. Drezner et al. 2001), it is much too abstract to be of real value to actual industrial location decisions (cf. Götze 1995, p. 56). A general criticism of Weber's theory can be found in Behrens (1971, pp. 15-19) and Meyer-Lindemann (1951, pp. 55-67).

Another important field of research has been to identify factors influencing location decisions in order to develop a system of location factors (cf. Table 1 for exemplary location factors). Frequently cited systems were published for example by Rüschenpöhler (1958) and Behrens (1971). Behrens grouped location factors into sourcing, transformation and sales factors. If one of the groups dominates a company's location decision this can be referred to as its *location orientation* (Behrens 1971, pp. 82-88). Government-oriented factors were later added in the context of international location decisions (cf. Bea 2000, pp. 341-342; Tesch 1980, pp. 359-519; Sabathil 1969, pp. 50-226). A review of the English literature on industrial location factors can be found in Blair and Premus (1987).

Table 1. Location factors[8]

Sourcing	Transformation	Sales	Government/ community
• Labor supply and skills • Labor costs • Raw material availability • Third party services • Industrial properties • Utility availability & costs • Transportation infrastructure • Waste disposal facilities	• Climatic suitability • Distance from housing areas • Agglomeration factors	• Market potential • Competition • Origin goodwill • Agglomeration factors • Decentralization factors	• Subsidies • Taxes • Legal system • Political stability • Labor policies • Trade barriers • Business climate • Cost and quality of living

Within the set of location factors it is common to distinguish between *quantitative* and *qualitative* factors, often without properly defining the difference between them. In the course of this work, factors that have a directly measurable financial impact will be referred to as quantitative factors and all other location factors will be considered of qualitative nature. A further distinction can be drawn between *qualifying* and *ranking* factors (cf. Pellerin et al. 2003, p. 268) with the former specifying minimum requirements and the latter being used to rank feasible alternatives.

Many empirical studies have been conducted on the relative importance of the various location factors (e.g., type of facility to be located, industry the facility belongs to and location region). MacCarthy and Atthirawong (2003) report findings of a Delphi study on factors affecting international plant location decisions, Grabow et al. (1995) specifically focus on qualitative factors, Brede (1971) examines the importance of a wide range of factors for various industries in Germany, Haigh (1989) provides a detailed discussion of location factors and decision processes employed based on a survey of 20 international companies with plant locations in the U.S., Lopez and Henderson (1989) examine location factors for food processing plants in the U.S., Artikis (1991) for the Greek food industry, Chernotsky

8 cf. Jung (2004), pp. 60-70; Haigh (1990), p. 27; Peskin and Halpern (1990); Wardrep (1985); Stafford (1980), pp. 44-155; Timmermann (1972), p. 390; Behrens (1971), pp. 47-81; Rüschenpöhler (1958), pp. 83-176.

(1983) analyzes the factors attracting German and Japanese companies to the Charlotte, N.C. area, Tong and Walter (1980) survey 254 foreign firms with manufacturing locations in the U.S. to assess the importance of 32 location factors and Bass et al. (1977) analyze 118 international plant location decisions of U.S. companies. Additional empirical analyses have been published by Brush et al. (1999) and Blair and Premus (1987). A recent empirical analysis for the U.S. that also contains references to other empirical studies on a factor by factor basis was provided by Karakaya and Canel (1998). Schmenner et al. (1987) analyzed the relative importance associated with various location factors at different stages of a location decision process. Other researchers have analyzed the influence of specific factors such as government incentives or environmental regulations on location decisions of companies. For example, Single and Kramer (1996) specifically investigate the effect of tax policy on plant location and provide references to research on other factors. Bankhofer (2001, p. 32) also lists publications on the relevance of specific location factors.

In addition to academic publications, publications directed primarily at practitioners faced with location decisions are available. Kinkel (2004), based on a review of academic publications, proposes various "instruments" for evaluating both national and international sites and provides application examples for each. The instruments include selection of location factors, a systematic collection of lessons learned from previous location decisions, scenario and options evaluation and location controlling. Hack (1999) provides a comprehensive step-by-step approach to site selection. He proposes a planning process including an evaluation process based on a simple scoring model. Various location factors typically important in site selection decisions are discussed in great detail and data sources are listed. Additionally, the appendix contains several checklists and survey questionnaires. Based on comprehensive empirical analyses Schmenner (1982b) provides guidance on how to conduct location decisions for industry but also gives advice to states and localities on how to attract industry.

2.3 Specialty Chemicals Production

2.3.1 Process Industries, Chemical Industry and Specialty Chemicals

Production systems can be separated into discrete parts production and process industry production. In discrete parts production, countable objects are modified or assembled in a sequence of production steps and mechanical production steps dominate. In process industry, substances are extracted, transformed, purified or mixed and chemical or biological processes dominate. In literature, no common definition of the characteristics of process industry production exists. Frequently cited attributes are multiple analytic and/or synthesis steps using raw materials of solid, liquid and gaseous state, cyclic material flows, creation of byproducts and variable yields. Industry segments included in the definition of process industry among others are chemical production, pharmaceutical production, food processing, paper production, petroleum processing and certain basic material extraction industries.[9]

The focus of this work is on chemical industry. Pharmaceutical production is included in the definition of chemical industry because the production technologies employed are very similar to those in chemical industry. Within chemical industry Kline (1976, pp. 110-113) suggests to define the four sub-segments also shown in Figure 7.

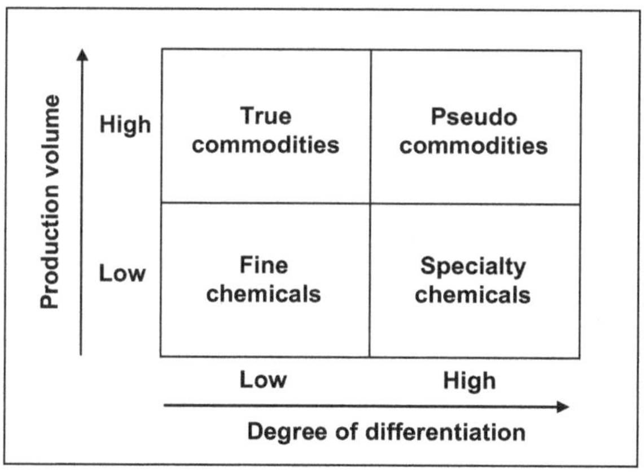

Fig. 7. Chemical industry segmentation

Kline (1976, pp. 110-111) and Amecke (1987, pp. 63-65) provide further characteristics and examples for each of the segments:

- *True commodities*, identified by their chemical structure, are usually produced by several suppliers in identical form based on a generally accepted standard. The value added is limited and raw materials are the

[9] This paragraph is based on Blömer (1999), pp. 5-9; Günther (1998), p. 356; Packowski (1996), pp. 33-39 and Corsten and May (1994), pp. 873-880 where further details on the topic can be found.

dominant cost factor. Typical examples are substances such as hydrochloric acid, hydrogen peroxide, ethylene glycol, etc.

- *Pseudo commodities* differ from true commodities in that they are not only defined by their chemical structure but that their application characteristics are optimized, too. The value added is also low and the share of raw material costs is high. However, for each product group a number of products with different application characteristics exists. Typical examples are fertilizers, solvents, elastomers, etc.
- *Fine chemicals* are also identified by their chemical structure. However, they usually have complex production processes and consequently a high value add. Typical examples are amino acids and pharmaceutical active ingredients but also highly concentrated forms of commodity chemicals.
- *Specialty chemicals* are developed to solve a certain application problem. Consequently, they form the largest of the four segments with respect to the number of products. Examples range from antifreeze compounds to pharmaceutical active ingredients.

The specialty chemicals market is very fragmented. As can be seen in Figure 8, it is common to distinguish between more than 30 primary and 350 secondary segments. This diversity is mirrored in today's industry structure where the ten largest companies have a market share of approximately 30% and a large number of very small players is active in the various niche markets (cf. Bartels et al. 2006, p. 96). Major specialty chemicals companies necessarily serve multiple segments within this market. As a consequence they operate a diverse production system with limited interdependencies between the different plants and thus have a relatively large degree of freedom in production network design (cf. also Chaps. 2.3.2 and 2.3.3). At the same time competitive pressures from low-cost market entrants and the price-cost squeeze caused by the combination of rising raw material prices and stagnant product prices require improvements of the cost base and make further industry consolidation likely (cf. Bartels et al. 2006, pp. 97-101). The need to support these processes with strategic (re-) designs of the production networks will be pronounced. For these reasons, this work focuses on specialty chemicals industry.

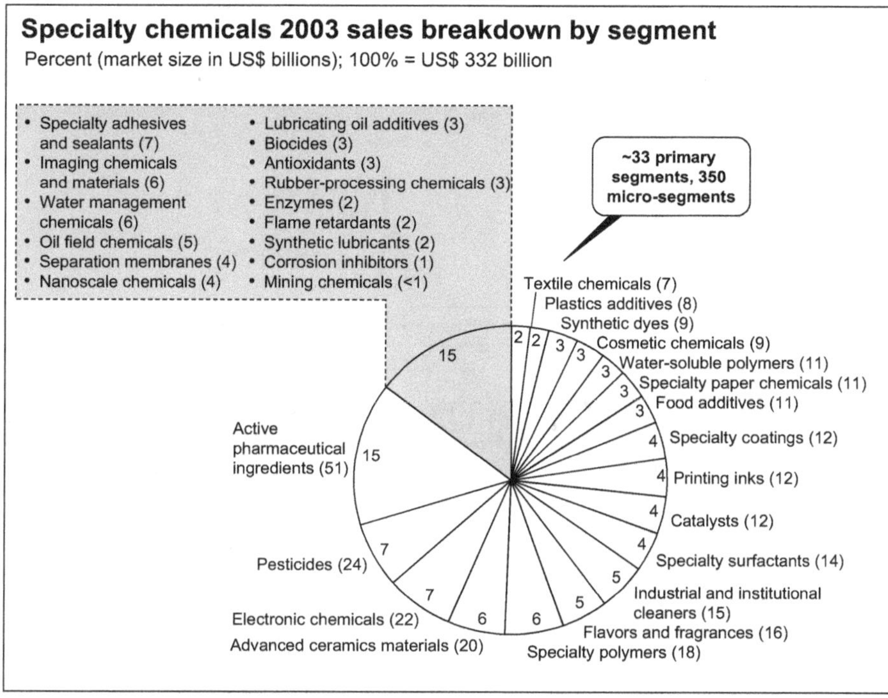

Fig. 8. Specialty chemicals market segments (source: Bartels et al. 2006, p. 96)

2.3.2 Chemical Production Sites

The terms production site, factory and plant are in many cases used synonymously but in the context of chemical production networks they have a distinct meaning. A *site/factory* is the geographical location where production takes place. Within the site, one or more *plants* produce different kinds of products. Additionally, a chemical production site contains several infrastructure units. In contrast to other industries, site infrastructure does not only comprise services such as security, internal logistics, maintenance, canteens and site administration but also utility plants and waste treatment facilities (cf. Fig. 9). For the 50 sites operated by the industrial cooperation partner the share of total costs associated with site infrastructure was in line with industry averages of 15% to 20% as reported by Rasch (2006, p. 257).

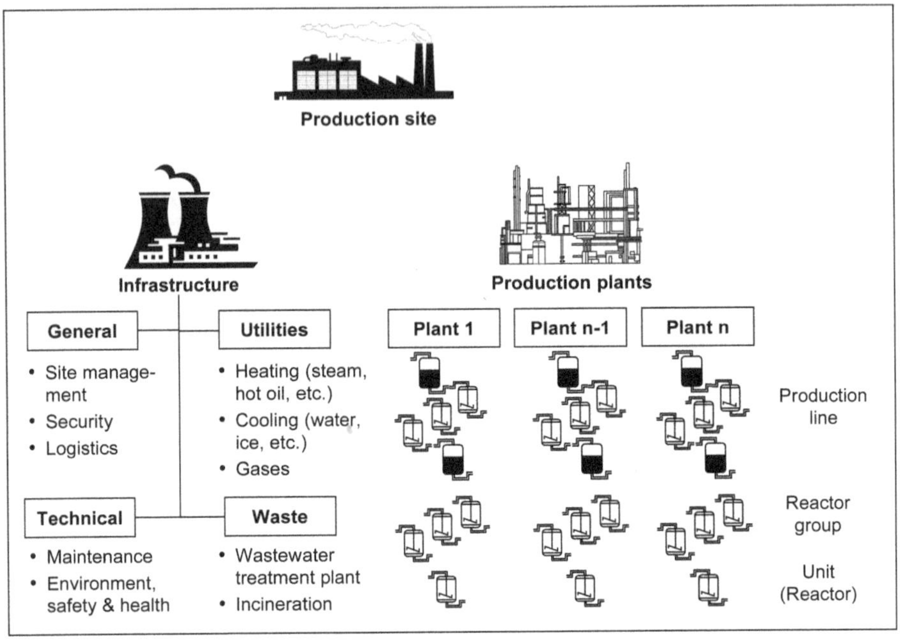

Fig. 9. Setup of chemical production sites

Site infrastructure commonly belongs to and is operated by the chemical company producing at the site. As the infrastructure costs are largely fixed costs, economies of scale can be achieved by concentrating production at larger sites. Recently, as a consequence of restructurings in chemical industry (e.g. the breakup of Hoechst into several companies, cf. N.N. 2005a), the model of operating sites as industrial parks has become more popular. A separate legal entity operates the site and provides infrastructure services to all production companies. This infrastructure company seeks lo locate other companies at the site to grow its own business and thus helps spread fixed costs. Another advantage of locating production at an industrial park is that for the individual company infrastructure costs can become largely variable costs. Additionally, setting up a new plant at an industrial park reduces the capital investment required in comparison to building a new site. Providing chemical industrial parks has also become a means of attracting investments (cf. Hoffman 1997). An exemplary industrial park, ChemSite in the Ruhr region of Germany, is profiled in N.N. (2003a). It should be noted that production managers operating plants in industrial parks also cite disadvantages such as long decision processes and a lack of cost-competitiveness of the services provided by the operating company.

A plant contains all production units required to produce a certain intermediate or finished product. The relationship between a site's plants varies between two extremes. At integrated sites (Verbundstandorte) the plants cover certain steps of the overall production process and are closely linked by material flows. This can be seen as a "plant within a plant" concept based on a process focus. Integrated sites are especially common in production of commodity chemicals. For example, the new integrated site built by BASF in Nanjing, China, consists of a steam cracker producing among others ethylene and propylene. Nine other plants further process the substances. The overall investment to build the site was U.S.$ 2.9 billion.[10]

Contrary to integrated sites, at specialty chemicals sites, material flows between the plants are usually limited or nonexistent. The structure of a typical production site can be seen as a "plant within a plant" concept based on a product focus where individual plants share site infrastructure to realize economies of scale. Only at larger sites a process-based separation can be observed, e.g., a central milling and packaging plant serving several pigments plants. Additionally, material flows are typically convergent and creation of by-products is an exception. Hence, individual plants can often be (re-) located without affecting complex material flow connections to other plants. As a consequence, the degree of freedom in designing production networks is much higher in fine and specialty chemicals than in commodity chemicals. Also, investment expenditures for a single specialty chemicals plant typically do not significantly exceed the amount of U.S. $ 100 million with many plants requiring much lower investments.[11]

Within a plant, equipment is often organized in production lines. Variants of the product family produced by a plant are clustered and assigned to different lines to reduce the complexity of changeovers (e.g., pigments production lines for different color shades within a family of chemically similar pigments). The organization of equipment below the level of production lines is not relevant in the context of production network design (Packowski (1996, pp. 129-131) provides more details on the subject).

2.3.3 Production Technologies in Chemical Industry[12]

Production network design includes decisions on production capacity and – if technically possible – choice of production technology. Consequently,

---

[10] For details on the Nanjing site see for example Roth (2005) or the BASF website at www.basf.de.

[11] cf. for example McPadden (2001); N.N. (2000).

[12] This chapter is based on Yang (2004, pp. 17-24), Blömer (1999, pp. 9-18) and Kölbel and Schulze (1967, pp. 15-36).

an understanding of different production technologies available is important. Plant types found in chemical industry offer different levels of product range flexibility:

- *Multi-purpose plants* combine different equipment units with flexible piping systems. They allow production of a broad range of products that vary considerably with respect to number and type of synthesis steps.
- *Multi-product plants* allow the production of a range of products with similar synthesis steps (usually a product group). Individual units are combined according to the production process of the product group.
- *Dedicated plants* are designed specifically for production of a single product or a small range of product variants. Redesigning dedicated plants to produce a different product is usually not possible.

Two basic types of material flow are common in chemical industry:

- *Continuous plants* have continuous raw material and product flows with a relatively constant transfer rate. Accordingly, the time required to produce a certain quantity is proportional to the production volume.
- *Batch plants* have discrete raw material and product flows. Equipment is charged before the process step and discharged after completion of the step. Process times are, within the limit of the equipment capacity, independent of the production volume.

The selection of the most appropriate plant type mainly depends on the underlying chemical process and planned production volumes. Continuous production is generally more economical than batch production. However, many production processes are not stable enough to be run in continuous plants as these offer fewer options to adjust the production process to variability in process performance. Typically, continuous plants are at the same time dedicated plants while dedicated batch plants are an exception. Therefore, continuous plants are chosen only if production volumes are expected to remain high enough to justify the investment into dedicated production technology. Multi-product plants can be found both in the form of continuous and batch plants. If batch equipment is used, it is usually operated in campaign mode where a sequence of several identical batches is produced before the equipment is prepared for the next product. Finally, multi-purpose plants practically always operate batch production equipment with changeovers taking place frequently (sometimes after every batch).

An additional lever often discussed in global production network design is to adapt production processes and sometimes also product design to the characteristics of the designated plant location (cf. Meyer 2005, pp. 119-

128). Factors typically playing a role in these considerations include factor cost differences, personnel skill levels and the expected production volumes. In chemical industry, the characteristics of the chemical process determine to a large extent the requirements of the production process. Location-specific choices are available primarily with respect to equipment capacity and degree of automation.

In specialty chemicals, the majority of plants are multi-product or multipurpose plants operated in batch mode. Only a relatively small proportion of products is produced in continuous production plants. This might change however, once current development activities ongoing in the area of flexible, continuous production technology for small production volumes (micro-reaction technology) find their way into industrial production systems.[13] So far, this technology is mostly employed in pilot-plant - scale production trials, but further improvements are to be expected in the near future.

The technical capacity of a plant is determined by the capacity of individual production lines/equipments and the number of parallel installations. Theoretically, a combination of production lines with different capacities at a plant allows for a wide range of choices in capacity selection. However, in practice operating only one production line size per plant is preferred because it is easier to switch products between lines if these are of similar size. While the capacity per production line can theoretically be chosen on a continuous scale often standard equipment sizes available from vendors lead to a discrete set of choices.

2.3.4 Specialty Chemicals Production Networks

Design Principles of Specialty Chemicals Production Networks

Specialty chemicals production networks can be looked at from two different perspectives: the *site perspective* and the *value chain perspective*. As discussed in Chapter 2.3.2 a specialty chemicals production site usually hosts several plants that may each belong to a different value chain. At the same time a value chain's plants are typically spread across several sites located in different economic regions. As a result one obtains a matrix structure such as the conceptual matrix shown in Figure 10. For each value chain the grey triangles show the importance of the different sites for the

[13] Microreaction technology is described in detail in Ehrfeld et al. (2000). Wille et al. (2004) describe an application in specialty chemicals.

respective value chain based on the share of global capacity located at each site. Analogous, for each site the white triangles indicate the relative importance of the different value chains located at the respective site as measured by the site fixed costs absorbed by this value chain.

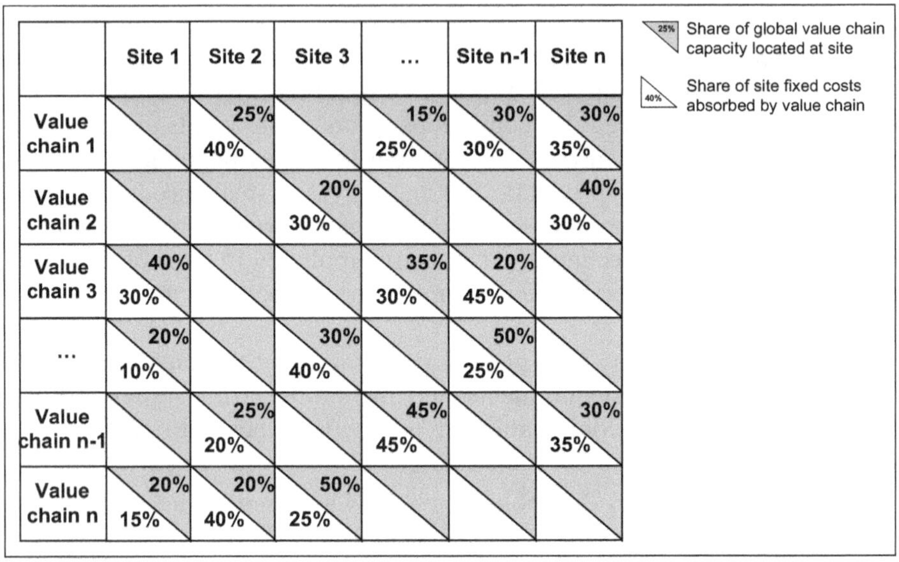

Fig. 10. Specialty chemicals production network matrix

Figure 10 also explains why any production network design decision in specialty chemicals needs to take into account the interdependencies between the two perspectives. Restructurings at the value chain level usually affect several production sites. The decision to relocate a value chain's production to a different site may for example endanger the cost competitiveness of the entire site if a large share of the site's fixed costs was absorbed by this value chain. Similarly, the decision to close a production site generally affects multiple value chains and may require network redesign projects for each value chain to decide on how to redistribute production.

Since the primary role of sites in specialty chemicals production networks is to provide infrastructure services for the different plants they host, the design principle underlying specialty chemicals production networks has to be analyzed from the value chain perspective. Both market area plant and product plant segmentations can be found (cf. Schmenner 1982a, p. 13). Often, in case of relatively small production volumes, each plant produces a segment of the value chain's product portfolio for global distribution. If production volumes are sufficiently high, a market area segmentation is also possible, whereby an identical product is produced at several plants for regional distribution. Process-based segmentations, combining similar process steps of different products/ product families in one facility, are typically not feasible in chemical industry. Additionally, process characteristics are often very similar across products with respect to the general nature of the inputs required.

Historically, the strategic role of foreign production sites explicitly established and not added in a merger or acquisition mainly was to access regional markets (server factories). One underlying reason were trade barriers such as tariffs which, according to Ferdows (1997b, p. 74), have declined from an average of 40% of product value in 1940 to 7% in 1990. With the increasing integration of global production networks, today many newly established sites are set up to exploit cheap factor costs (offshore or source sites). A recent example is the decision of Degussa to shift a large proportion of its pharmaceutical production from Europe to China (cf. N.N. 2004). If markets require local research & development activities, contributor factories take over responsibility for certain products or development tasks. Lead factories are commonly located in the home country of the respective company with foreign lead factories mostly added in the course of mergers & acquisitions.

A Real-World Specialty Chemicals Production Network

Today's specialty chemicals industry consists of several companies with global operations and countless regional/niche competitors. To illustrate the complexity of specialty chemicals production networks, the network operated by American company Rohm and Haas is presented here. It is chosen because of the public availability of data[14] but is representative of other specialty chemicals companies' networks.

Rohm and Haas was originally founded in Germany in 1907. A branch established in the U.S. in 1909 became independent in the wake of World War I. The company was active in several countries before World War II. Its most renowned product, Plexiglas, was also developed at that time. Rohm and Haas underwent both reorganizations and product portfolio changes in the period until 1990. During this time further investments into international operations were undertaken as well. In the decade from 1990 to 2000 participation in the mergers & acquisitions activities ongoing in specialty chemicals industry lead to a significant reshaping of the product

[14] The information on Rohm and Haas' production network is taken from the 2005 annual report (cf. Rohm and Haas (2006), pp. 36-37).

portfolio and the operational footprint. In 1992, Rohm and Haas acquired electronics chemicals company Shipley. In 1997, a minority interest in Rodel, Inc. was bought. In 1999, specialty additives company LeaRonal, Inc. was bought and the company merged with Morton International creating today's company with sales of US$ 7,994 million in 2005.[15]

Fig. 11. Rohm and Haas' global production network

Rohm and Haas is organized into six operating segments: coatings, adhesives and sealants, electronic materials, performance chemicals, salt and monomers. According to its 2004 annual report, Rohm and Haas operates approximately 100 manufacturing facilities and 35 research and technology facilities in 27 countries worldwide. Of the 95 manufacturing locations listed explicitly, 41 are located in the United States, 24 in Europe, 17 in Asia (including Australia and New Zealand), 7 in Canada, 3 in South America, 2 in Mexico and 1 in Africa. While 59 sites host only operations from a single business segment (information on the number of plants per site is not available), the remaining 36 sites contain operations from up to four different business segments. Figure 11 shows the global production network.

[15] This paragraph is based on the company history published on www.rohmhaas.com.

2.4 Production Network Planning and Controlling

2.4.1 Production Network Planning Process

Many authors dealing with production network design or plant location do not discuss the planning process at all but instead focus solely on decision support models. If the planning process is discussed, it is often in the context of locating a single plant excluding network design considerations. Issues arising from the more common case of redesigning existing networks are covered even less frequently. Below, the publications that form the basis for the planning process proposed here are briefly reviewed.

The planning processes found in literature that focus on locating a single site generally disaggregate the problem into several levels. Bankhofer (2001, pp. 116-118) suggests a two-step approach starting with the selection of a macro-location (region or country) followed by the selection of a micro-location. Within each step, theoretically feasible alternatives are pre-selected (scanning). For those alternatives that have passed the screening, conformity with minimum requirements is checked before performing a more detailed analysis. To facilitate the detailed analysis, Bankhofer discusses various qualitative and quantitative methods from literature and develops a new approach based on multivariate statistical analysis. Similar to Bankhofer, Autschbach (1997, pp. 193-222) divides the evaluation phase into country pre-selection, rough analysis and detailed analysis of alternatives. While he mentions different evaluation methods, Autschbach does not suggest the use of any specific tool. Goette (1994, pp. 254-319) suggests a process consisting of five phases: concept development, country pre-selection, macro-analysis, micro-analysis and decision. He suggests using qualitative tools such as checklists, profiling method or scoring models to evaluate alternatives within the different analysis steps.

Götze (1995, pp. 63-185) develops a planning process that explicitly includes the design of the overall production network. He suggests a hierarchical planning process starting with the development of a production network design strategy followed by a site selection phase (if required), a detailed network design phase and finally an implementation planning. The end product of the network design strategy phase is a network blueprint including the number of sites, their rough geographical distribution and the general principle for allocation of products and resources to the sites (cf. Götze 1995, p. 128). The optional site selection phase is further sub-divided into a macro phase for country selection and a micro phase for selection of the actual site. Götze discusses various decision support tools for the different phases of the planning process (cf. Götze 1995, pp. 221-408). For network design, he presents three optimization models: a transportation cost minimization model assuming a single product, a model for allocating several products (both finished goods and intermediates) to a set of existing sites and an integrated network design model for simultaneous location and allocation decisions. To make using the integrated model feasible, Götze suggests to use only very aggregated data and combine it with other decision support tools for more detailed analyses. For site selection, he proposes to combine a scoring model and a dynamic investment appraisal calculation. While Götze covers the full range of issues arising in production network (re-)design the main drawback of his approach is the assumption that it is possible to perform (rough) network design and detailed product and production capacity allocation in separate planning steps effectively ignoring existing interdependencies.

The planning process proposed by MacCormack et al. (1994) also explicitly includes network design. They propose a four-phase process starting with establishing the role required of manufacturing in the context of the respective business environment. In a second step, market access, risk management issues, customer demand characteristics and the impact of production technologies on plant scales are analyzed to develop a production network configuration that includes the number of sites per region, product allocation and capacities. In a third step, a set of potential sites is identified primarily based on infrastructure requirements of the business and the underlying manufacturing strategy. In the last phase, MacCormack et al. recommend the use of "computer-based models" to determine the detailed configuration of the actual production network within the boundaries imposed by the previously developed rough network design.

None of the existing planning processes covers the full range of issues and especially the complex interdependencies inherent in specialty chemicals production network design. To close this gap, a planning process that is applicable to the design of new production networks and the redesign of existing networks is developed below.

One element all of the existing planning processes have in common is that the problem is decomposed to reduce the complexity of the planning task. However, neither a sequential planning process (number of plants followed by location, capacity and technology) nor a geographical (selection of economic region followed by country and site) or hierarchical decomposition (rough network design strategy and detailed network planning) can cover all interdependencies. As Günther and Tempelmeier (2005, pp. 68-70) point out, different levels of site selection/network design decisions are dominated by structurally different types of location factors.[16] Figure 12 shows the four levels proposed by Günther and Tempelmeier and major location factors relevant at each level. At the level of economic region and country (large countries may have to be sub-divided), quantitative factors such as market potential, transportation costs, factor cost differences and trade barriers dominate the decision. Selection of a province and physical site are mainly driven by qualitative factors such as infrastructure and raw material availability, but, to varying degrees, local factor cost differences and site-specific property/investment requirements also affect the decision.

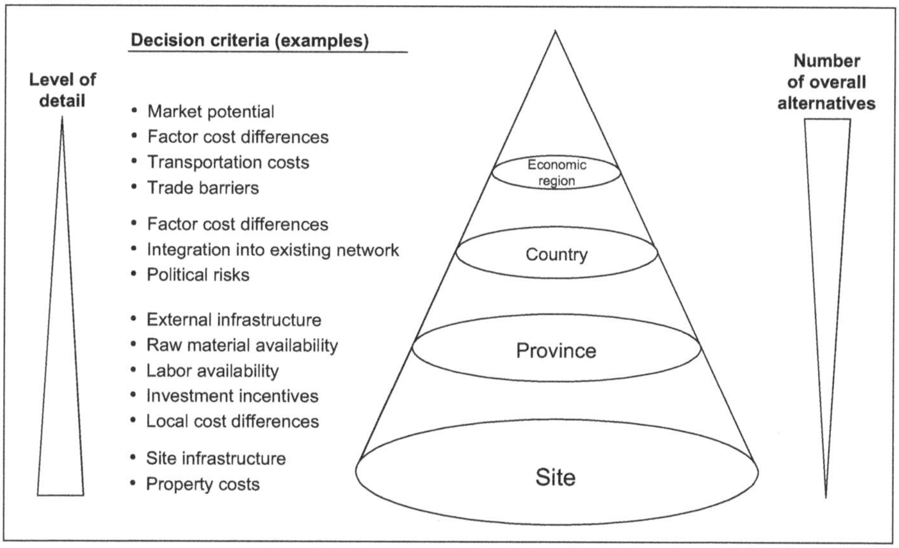

Fig. 12. Production network design decision levels

Considering a global solution space, this structural difference of location factors can be exploited to decompose the planning process into a network design phase at the country level, supported by quantitative decision support tools and a site selection phase within individual countries that is supported by a qualitative decision support tool. In contrast to the planning process proposed by Götze, the network design phase simultaneously determines number, location, capacity and product/market allocation of the sites forming the network. The exact location of a plant within a country is determined in the site selection phase taking into account the previously

[16] Schmenner et al. (1987) explicitly study the relative importance of location factors at the state and actual location level.

defined role within the global network (e.g., transportation infrastructure requirements for an export site).

		Production network optimization			Site selection		
	Problem definition	Identification of alternatives	Evaluation of alternatives	Decision	Identification of alternatives	Evaluation of alternatives	Decision
Activities	• Initiation of network planning process • Definition of objectives • Analysis of interdependencies • Definition of scope and restrictions	• Definition of location factors and data collection • Model development • Identification of alternative network configurations • Quantification of economic impact of restrictions	• Sensitivity analyses • Cost-benefit analysis • Risk analyses	• Selection of preferred network design alternative	• Identification of location factors • Pre-selection of feasible alternatives • Collection of detailed data for feasible alternatives	• Systematic evaluation of remaining alternatives • Sensitivity analyses • Cost-benefit analysis	• Selection of preferred location
End products	• Project charter	• Set of alternative design options	• Rank-ordered alternatives		• Set of 5 to 10 alternative sites	• Rank-ordered set of sites	

Fig. 13. Production network design planning process

The proposed planning process starts with a problem definition phase, followed by the network optimization phase and the site selection phase. As shown in Figure 13, both the network design and the site selection phase consist of three sub-phases: identification of alternatives, evaluation of alternatives and decision.[17] To which extent these phases are required and whether they have to be further sub-divided has to be determined within the problem definition phase based on the specific planning task. Generally, the site selection phase is of minor importance for specialty chemicals companies that already operate a global production network. They instead prefer to locate new plants at existing sites to realize economies of scale and limit the complexity of the production system. The same holds true for other industries as empirical research published by Bankhofer and Kugler (1999, p. 28) demonstrates.

[17] The planning process is based on the general planning process as proposed by Hahn and Hungenberg (2001, pp. 45-47) consisting of the four phases problem definition, identification of alternatives, evaluation of alternatives and decision. Vos and Akkermans (1996, p. 64) employ a similar planning process for production network design and Mintzberg et al. (1976) find that similarly structured decision processes are typically employed in strategic decision situations.

In the following chapters the individual phases of the planning process and the major activities taking place in each phase are outlined. The decision support tools proposed for the production network optimization phase and the site selection phase will be developed in Chapters 3 and 4 respectively.

2.4.2 Problem Definition Phase

Initiation of a Production Network Planning Project

The need to set up a production network planning project can arise for many reasons. Bankhofer (2001, pp. 86-93) differentiates between causes requiring a production network (re-)design and causes requiring actions only at an individual site. However, his segmentation of causes is neither mutually exclusive, nor does it capture the interdependencies between site and value chain perspective typical of specialty chemicals production networks. Here, a segmentation based on the origin of the respective cause is proposed. Table 2 contains a list of typical causes requiring the initiation of a production network planning project.

Table 2. Causes requiring initiation of a production network design project[18]

Value chain	Site
• Increase or decrease of required capacity • Market access • Product portfolio changes • Lack of cost competitiveness • Trade barriers • Currency fluctuations • Changes in geographic distribution of demand • Integration of acquired companies	• Lack of cost competitiveness on infrastructure and/or production level • Lack of long-term development potential • Outdated buildings / production equipment • Changes in government regulations (e.g., environmental regulations) • Availability of resources / external infrastructure • Political risks

While the cause leading to the initiation of a project can be ascribed either to the situation at a single site or that of an entire value chain, the resulting project can usually not be limited to one of these perspectives. For

[18] cf. Günther and Tempelmeier (2005), pp. 67-68; Bankhofer (2000), p. 339; Emmrich (1997), pp. 245-247; Lüder and Küpper (1983), pp. 97-98 and 139-151.

example, if a site is to be closed, a production network redesign project might be required for all value chains affected. To analyze possible interdependencies and define the project scope, in a first step the data leading to a matrix similar to the one shown in Figure 10 on page 32 should be collected. A site's share of a value chain's overall capacity and vice versa a value chain's share of a site's cost base are good indicators for prioritizing the analysis of interdependencies. The extent to which other sites/value chains will actually be affected by the project depends on the design alternative selected and thus will become apparent only in the course of the project.

Objectives of Production Network Planning Projects

Generally, any production network planning project's primary objective is to increase capacity, reduce capacity or restructure existing capacity (cf. Schill 1990, p. 47; Lüder and Küpper 1983, p. 166). In many cases, production network planning projects have to consider additional, secondary objectives. Following the categorization used by Hahn (1999, pp. 37-38), these are either business targets, financial targets or social targets. Table 3 lists a number of objectives relevant to production network design. The secondary objectives can take the form of restrictions that have to be considered within the project (e.g., limits on restructuring costs) or be used to reduce the solution space (e.g., short lead time requirements may make overseas production impossible or the objective of reducing the complexity of the overall production system excludes adding new sites). While some of the secondary objectives are applicable to all production networks, many of them have to be specifically derived from the value chain's respective business strategy.

Table 3. Secondary objectives of production network planning projects[19]

Business targets	Financial targets	Social targets
• Closeness to customers • Access to know-how / R&D facilities • Reduction of production system complexity • Product quality • Access to new markets • Lead times • Flexibility • Access to raw materials	• Profit maximization or cost minimization • Minimization of cash-flow / EBIT impact • Operational hedging against currency risks • Utilization of subsidies	• Corporate image • Employee satisfaction • Environmental protection • Minimization of job losses

Types of Production Network Planning Projects

Project approach and solution space vary depending on the cause leading to the project initiation and the objectives of the project. Figure 14 shows different types of network planning projects. The design of a new production network, though technically a capacity expansion, is considered separately to account for the higher level of freedom in a greenfield approach compared to the redesign of existing networks.

If a capacity expansion is the primary objective, the two most common cases are

- The additional capacity needed does not require construction of one or more new plant(s) but can be achieved by expanding existing plants. Consequently, the solution space is limited to the value chain's existing production network and the plant most suitable for expansion has to be identified. In this case the plant to expand can often be selected based on investment appraisal calculations without a complete network design project.
- The additional capacity requires construction of a new plant. In this case location alternatives both within the value chain's existing network and outside this network have to be considered. Thus, a network design phase is required in all cases and a site selection phase if alternatives outside existing sites are to be considered.

[19] cf. Hagedorn (1994), pp. 27-30; Hummel (1997), pp. 85-95.

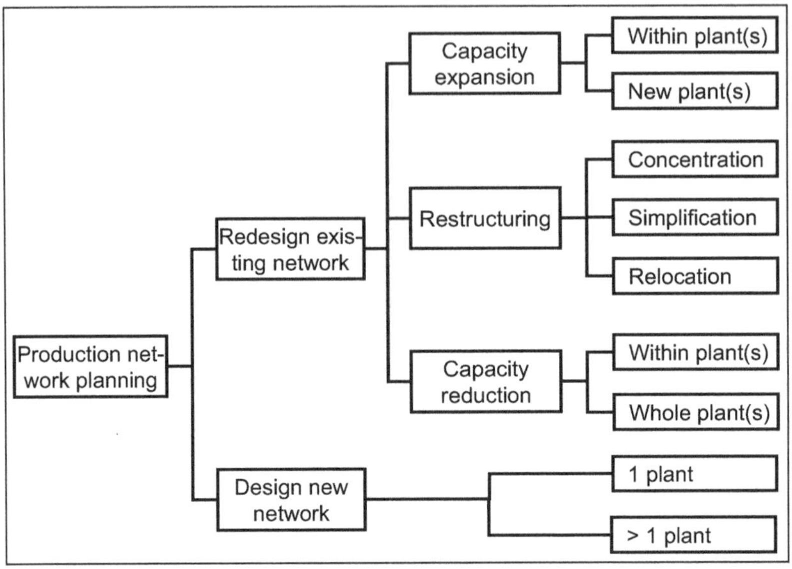

Fig. 14. Types of production network planning projects[20]

Similarly, if a capacity reduction is the primary objective, depending on the desired extent of capacity reduction, a plant can either be downscaled or closed entirely. In both cases usually neither network design nor site selection planning are required as the plant to be downscaled or closed is usually selected based on technical/financial considerations. Nevertheless, the closure of one plant can severely worsen the competitive position of other plants located at the site affected due to the fact that the remaining plants have to carry an increased fixed cost burden. As a consequence, a site closure project requiring network design projects for the other value chains present at the site might have to be initiated.

If the primary objective is the restructuring of a value chain, the solution space is wider as several plants can be affected simultaneously. Restructurings are usually initiated because of inefficiencies within the existing network or in the context of site closures and can take on one of three basic forms (cf. Bankhofer 2001, p. 95):

- A *concentration* of the production network reduces the number of sites where a product is produced effectively reducing the complexity of the value chain's production network. Concentration can take place either at sites that are already part of the respective production network or can

[20] Adapted from Bankhofer (2001), p. 96.

entail construction of a plant at a new site. If a new plant is to be constructed, a site selection phase might be required in addition to the network design phase.

- A *simplification* of the production network redistributes product variants within the network to reduce the number of plants producing identical products. This leads to a reduction of the network's complexity and allows for economies of scale. Usually, no change of capacity and no change in the number and location of plants being part of the value chain's production network takes place.
- A *relocation* of production transfers production to a more suitable site where usually a new plant has to be built. Even if the value chain's other plants are not included in the project scope, locating the new plant requires a network design phase. A site selection phase might also be required if the new plant is not to be built at an existing site or several existing sites have to be considered.

It should be noted though, that in practice the restructuring of a value chain's production network often combines different basic restructuring types and includes capacity changes. Also, the relative importance of decision criteria depends on the type of project. For example, in plant closures factors such as public opinion towards the decision (sometimes causing a company to favor closures in countries other than their home country) and possible obligations to repay subsidies received become important (cf. Richbell and Watts 2000).

2.4.3 Production Network Optimization Phase

Identification of Alternatives

As explained above, production network design has to simultaneously define the number of plants, their locations, capacities and technology. Additionally, product-plant and plant-market assignments have to be made. Due to the complexity of this task, it is difficult to first "collect" feasible network configuration alternatives and perform a detailed evaluation of these alternatives in a second step. Instead, a mathematical model of the production network can be used to generate alternative options for further evaluation (cf. Goetschalckx and Fleischmann 2005, p. 120; Vidal and Goetschalckx 1996, p. 10). Actually creating such a model has the additional benefit of improving the understanding of the supply network throughout the organization (cf. Geoffrion and Powers 1995, p. 106). As optimization models (alternatively, simulation models or heuristics could be employed) are considered to be the most suitable decision support tool for production network design (cf. Harrison 2003, pp. 7-8), the remainder assumes development of such a model. Since the actual model will be developed in Chap. 3 this chapter only contains a brief overview of issues.

The first step in developing a model of the production network is to define the factors that influence the network design decision and collect the required data. In practice, as Harrison (2003, p. 8) points out, collecting this data is often the most challenging aspect of a network design project. Many different functions of the company (marketing, production, sourcing, logistics, etc.) are affected and some of the data might not be readily available as actual values - not to speak of long-term forecasts in line with the time horizon of strategic network planning.

In the second step the actual model has to be developed. The model can then be used to generate alternative network configuration options. While the solution space at this stage theoretically includes every country, a subset of especially attractive countries is commonly pre-selected. Goetschalckx and Fleischmann (2005, p. 129) propose to generate alternatives by solving the optimization problem for differently specified objectives and alternative scenarios. Additionally, they suggest that "intuition and managerial insight" should be used to generate further alternatives. However, to be capable of performing these analyses, the data for all potential investment options has to be compiled. Since it is impossible to generate the required data for all available options, a sequential approach must be employed. For example, Stobaugh (1969) proposes to combine core country-related and product-related location factors to screen investment options and arrive at a shortlist. For this shortlist the required data can then be compiled and detailed evaluations be performed. In practice, this approach is typically associated with the risk of excluding attractive options without a proper analysis. This might also explain the finding from Meyer (2005, p. 13) that companies on average considered only 2.5 potential sites when analyzing relocation options.

Evaluation of Alternatives

As each alternative network configuration already possesses an optimum objective function value based on the scenarios/restrictions underlying the calculation, evaluation of the alternatives has to focus on aspects other than financial optimization.

One aspect becoming increasingly important is the robustness of the network configuration (cf. Goetschalckx and Fleischmann 2005, p. 118). One way of establishing a robust solution is to perform sensitivity or scenario analyses on major influencing factors such as demand expectations to determine how quickly the configuration analyzed becomes suboptimal in case of parameter variations. Alternatively, robust optimization techniques can be employed. The details of robust production network design will be discussed in Chapter 3.4.4. Additionally, a risk analysis of the resulting network configuration should be made. Topics of such an analysis could be to what extent the production network depends on an individual plant, how much of the capacity is located in countries with high political risks, etc. With restructurings the alternatives are associated with different levels of investments required and implementation risks from disruptions of ongoing production. An implementation risk/reward analysis of the options available can provide valuable insight into the question of whether the additional benefits outweigh expected implementation risks.

Decision

Once the alternatives have been evaluated, management has to chose the preferred network design and approve an implementation project. Alternatively, the planning process has to be repeated to develop further alternatives.

2.4.4 Site Selection Phase

The site selection phase is only part of the planning process if the network design selected in the previous phase requires the selection of a new site for construction or the choice between several existing sites for expansion, closure or relocation. If a new plant is to be constructed in a country where the company already operates one or more sites, as previously discussed it will usually choose to locate the new plant at an existing site to realize economies of scale. A detailed model to support site selection will be developed in Chapter 4.

Identification of Alternatives

The network design phase already determined the countries where plants should be located or closed. Thus, site selection takes place within an individual country. As the location factors pertinent to the site selection phase are different from those used in the network design phase, the first step again is to establish the relevant location factors. These are mostly of qualitative nature but also include quantitative factors such as local factor cost differences, property and construction costs. Findings from industrial location science (cf. Chap. 2.2.2) can be used as a starting point to define the location factors, but industry-, company- and project-specific factors have to be included, too. If the objective is to select a plant for closure, decision criteria mostly are the same but additional factors relevant in closure decisions such as the existence of environmental contaminations have to be considered.

In case of selecting a new site, a pre-selection of provinces based on minimum requirements for major location factors helps reduce the solution space at the beginning of the project. For example, if the plant is supposed to export a large proportion of its production volume, a location within a province that has an international port can be a pre-selection criterion. Within pre-selected provinces, alternative sites have to be identified. In addition to internal know-how, external sources such as government development agencies, chambers of commerce, trade associations, trade fairs and specialized consulting services can be used to identify feasible alternatives. Hummel (1997, pp. 201-221) and Autschbach (1997, pp. 157-165) discuss the suitability of various internal and external data sources. In case of sufficiently large investments it is also possible to publish a request for proposals placing the burden of preparing required site data on communities interested in hosting the site.

For new sites alternative implementation modes have to be considered, too. Generally, it is possible to acquire an existing site (or the company possessing the site), seek a joint venture partner to establish the site or build the site in a greenfield approach. The choice between these alternatives depends on numerous factors such as prior experiences in the selected country, sharing of know-how, government regulations, etc. For example, in certain industries China limits the equity share a foreign company is allowed to own making joint ventures mandatory (cf. Li and Clarke-Hill 2004, p. 59). An overview on foreign direct investment restrictions in OECD countries can be found in N.N. (2003). Autschbach (1997, pp. 67-80) discusses advantages, risks and motivations for the different options and Herzfeld (1983) covers the various aspects arising in joint venture investments.

If the number of alternatives identified is too large to perform a detailed evaluation for each, a pre-selection step should reduce the number of alternatives to 5-10 sites. A practical approach commonly used is a checklist approach to eliminate sites that do not meet a certain level defined for important location factors (cf. Wardrep 1985, p. 10).

Evaluation of Alternatives

The evaluation process has to lead to a rank ordering of the alternative sites and a recommendation of where to locate the plant. Evaluation methods proposed in the context of site selection can broadly be grouped into optimization, Multiple Criteria Decision Analysis (MCDA), simulation and investment appraisal calculations (cf. Hummel 1997, pp. 241-293). Götze (1995, pp. 335-339) proposes to combine an MCDA tool with an investment appraisal calculation to account for cost differences between the alternatives. This approach is in line with results from empirical research on site planning Bankhofer conducted with 53 large German industrial companies. He found that 86.8% use investment appraisal calculations, 81.1% MCDA tools, 28.3% optimization tools and 3.8% simulation approaches with more than 50% combining an MCDA tool and investment appraisal calculations (cf. Bankhofer 2000, p. 342).

Decision

Once the alternatives have been evaluated, again management has to choose the preferred site and approve the implementation project.

2.4.5 Integration of Production Network Design into Strategic Planning

As Liebmann (1971, p. 13) and Hummeltenberg (1981, p. 29) point out, strong interdependencies within the production network and between the production network and other functions such as research & development, marketing, distribution and finance require a holistic planning approach. Consequently, as many authors dealing with plant location such as Schill (1990, p. 3) and Götze (2002, col. 1827) demand, production network design and controlling have to be an integral part of a company's overall planning and controlling system. As specified by Lüder (1982, p. 430) production network strategy should comprise a long-term development strategy for each site and each value chain's production network.

In industry, as Schill (1990, p. 2) observes, production network decisions often appear to be considered only in case of obvious deficiencies requiring immediate action. One underlying reason may be the phenomenon of cumulative inertia (cf. Huff et al. 1992, pp. 56-57). In addition, empirical research from Lüder and Küpper (1982, p. 27) showed that production network planning was often only insufficiently integrated into the overall strategic planning process. More recent research from Bankhofer and Kugler (1999, p. 18) suggests that in the meantime at least a majority of companies (69.2% of companies polled) implemented site controlling processes.

To support an early identification of competitive threats requiring a network re-design Meyer and Jacob (2006, pp. 151-158) have suggested to integrate five indicators into controlling processes:

- *Geographic demand distribution*: changes in the geographical distribution of demand/demand growth rates or the geographical structure of orders received indicate a threat to the current network structure.
- *Education levels, technology transfers and cluster creation:* changes of a location's education level, technology transfers of competitors demonstrating the feasibility to move production to a certain location or the emergence of regional clusters within an industry indicate the future attractiveness of potential locations.
- *Geographic distribution of revenues and costs:* imbalances generally indicate that risks from exchange rate exposure might exist. Additionally, revenues in low-cost countries originating from sale of products manufactured in high-cost countries might not be sustainable in the long run.
- *Network redesign in industry:* network redesign activities in industry indicate the feasibility of restructurings but also the threat of increased competition due to an improved cost position achieved by competitors.
- *Increased imports from low-cost countries and price erosion:* Increased imports from low-cost countries are typically correlated with future price erosions indicating a strong increase of competitive pressure.

For specialty chemicals production networks, the planning and controlling process has to combine the site perspective and the value chain perspective. Most specialty chemicals companies have a division/business unit organization and operate sites shared by several divisions. While each division can cover the value chain perspective individually, for example by integrating the above-mentioned indicators into controlling processes, the site perspective requires cross-divisional coordination to avoid suboptimal solutions. Different options ranging from centralized to decentralized set-ups exist to do so (cf. Hayes and Wheelwright 1984, pp. 120-125).

Figure 15 shows how production network design and site strategy can be integrated into the strategic planning process of a company with a divisional organization structure. Production network design and capacity planning are integrated into each division's strategic production planning process. A cross-divisional site strategy team is established to coordinate production network projects and develop a consistent development strategy for each site. To provide the data required for strategic planning and identify inefficiencies as early as possible, production network and site controlling are integrated into regular performance reviews.

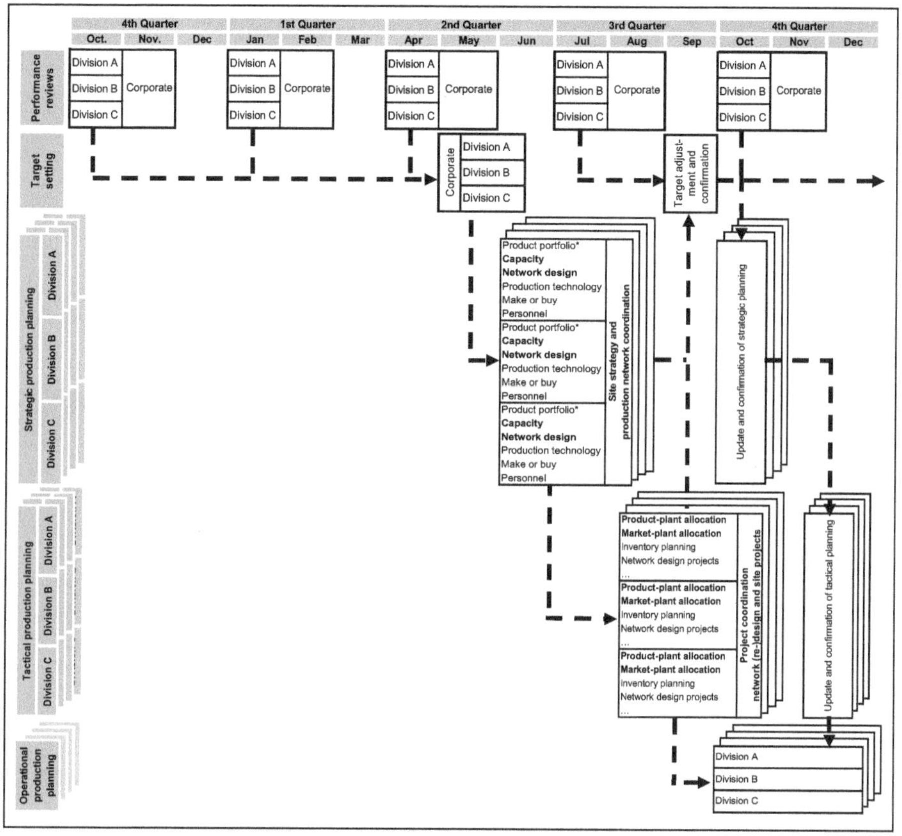

Fig. 15. Integration into planning and controlling processes[21]

[21] Source: Hahn and Hungenberg (2001), pp. 806-807 with modifications by the author.

3 Global Production Network Optimization

The production network design process developed in Chapter 2.4 is split into a global network optimization phase at the country level and a site selection phase focusing on the evaluation of individual production sites within a country. In this chapter the mathematical optimization model required to support the global network optimization phase of specialty chemicals production networks is developed. To this end Chapter 3.1 briefly introduces general research in the field of quantitative location analysis. The literature specifically focusing on production network design is reviewed in greater detail in Chapter 3.2. Alternative approaches to model major elements of production networks and their applicability to specialty chemicals industry are discussed in Chapter 3.3 and tailored modeling approaches are developed. Chapter 3.4 contains the resulting Mixed-Integer Linear Programming (MILP) model and possible extensions to include features that were not required in the pilot application underlying this work (cf. Chap. 5). Finally, Chapter 3.5 presents the results of numerical performance tests to demonstrate the applicability of the model to problem instances of realistic size.

3.1 Location Analysis and Production Network Optimization

ReVelle (2001, p. 459) defines location analysis as "the development of formulations and algorithms/methodologies to site facilities of diverse kinds in a spatial or geographic environment". In the following, the facilities to be located are further specified to possess point characteristics excluding models dealing with the location of area objects. The latter are employed to support facility layout and hence are not relevant in the context of this work (cf. Domschke and Krispin (1997) or Francis et al. (1992) for further references on facility layout models).

Figure 16 contains the most important criteria to classify facility location problems. Regarding solution space, as Francis et al. (1983, pp. 221, 240) explain, discrete location models are the most realistic (especially be-

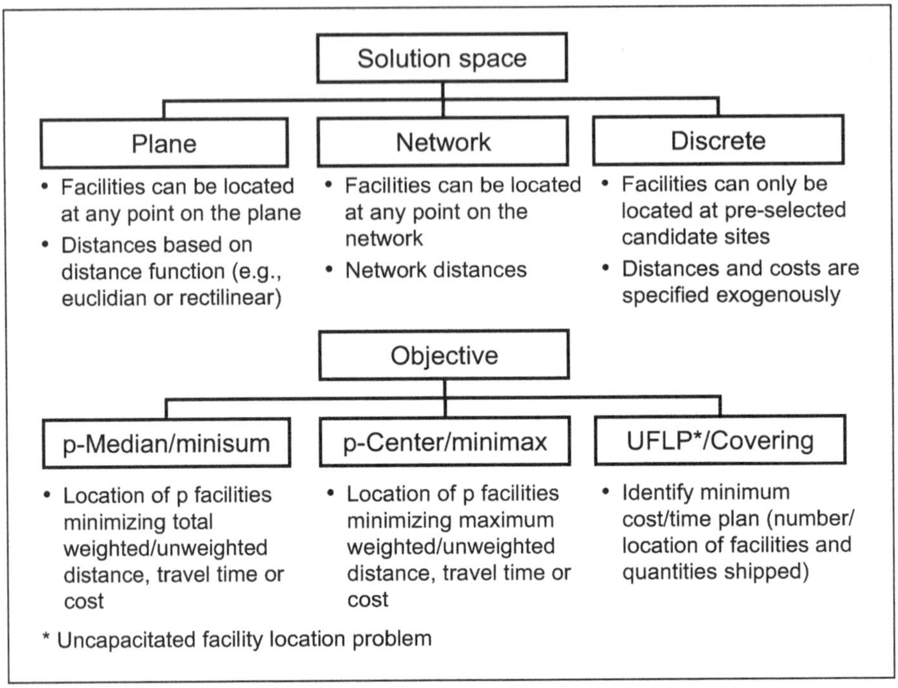

Fig. 16. Classification of facility location problems[22]

cause location-specific cost differences can be modeled) but also the most demanding in terms of data requirements and computational complexity. The characteristic difference between p-median/p-center problems and the uncapacitated facility location problem (UFLP)/covering[23] problem is that with the former the number of facilities to be located is specified exogenously while the latter also determines the optimum number of facilities. The classical Weber problem (cf. Chap. 2.2.2) is a 1-median location problem on the plane. P-center network models are for example used to locate emergency services such as police stations on a street network. In case of locating undesirable facilities the objectives can also be inverted, e.g., to maximize the minimum distance (cf. Erkut and Neuman 1989; Melachrinoudis and Cullinane 1986). Also, the competitive environment might be considered (cf. Plastria 2001; Bauer and Domscke 1993; Ghosh and Harche 1993) or multiple objectives be pursued simultaneously (cf. Current et

[22] cf. Krarup and Pruzan (1990); Francis et al. (1983).

[23] Covering problems are a special case of the UFLP where the objective is to identify the minimum number of facilities required to cover demand with a maximum cost/time/distance constraint.

al. 1990). Production network design models typically are specifications of the UFLP with a discrete solution space.

Location analysis has received considerable attention in the operations research community and hence the literature on facility location is extensive. However, the classical facility location literature is too general to be a good starting point to develop a model for specialty chemicals production network design and hence a review is not within the focus of this work. For reviews the reader is instead referred to ReVelle and Eiselt (2005), Hale and Moberg (2003), Owen and Daskin (1998), Domschke and Krispin (1997), Sridharan (1995), Brandeau and Chiu (1989), Francis et al. (1983) and Tansel et al. (1983a and 1983b). Detailed discussions of facility location issues can also be found in the books written/edited by Drezner and Hamacher (2001), Drezner (1995), Francis et al. (1992) and Mirchandani and Francis (1990). Eiselt (1992) reviews publications focusing on practical applications and Ghosh and Harche (1993) focus on private sector applications and different types of competitive models.

3.2 Review of Supply Network Optimization Literature

Comprehensive reviews of the literature on supply network optimization models have been published amongst others by Daskin et al. (2005), Melo et al. (2005), Bhutta (2004); Geunes and Pardalos (2003), Goetschalckx et al. (2002), Tsiakis et al. (2001), Goetschalckx (2000), Schmidt and Wilhelm (2000), Beamon (1998), Vidal and Goetschalckx (1997), Thomas and Griffin (1996) and Verter and Dincer (1995 and 1992). In face of the availability of the above-cited reviews, it is not intended to prepare another comprehensive literature review. Instead, the objectives of this chapter are twofold. Firstly, a classification of models from literature is provided since the above-cited reviews do not use a uniform approach towards classifying the models they review. Secondly, selected models are reviewed in greater detail with the selection criterion being either the application-oriented nature of the model or the novelty of the publication (i.e., publications that are not included in one of the reviews cited above). Additionally, Chapter 3.3 contains comprehensive reviews of alternative modeling approaches for major aspects of production networks as a basis for developing formulations suitable for specialty chemicals industry.

3.2.1 Classification of Supply Network Optimization Models

Categorization schemes have been suggested both for facility location (e.g., Hamacher and Nickel 1998; Ballou 1992, pp. 323-324; Brandeau and Chiu 1989, pp. 647-650) and supply chain optimization models (e.g., Bankhofer 2003, pp. 27-34; Bestmann 2001, pp. 46-47) and many literature reviews contain classifications of the models they review. The following criteria (the abbreviations in brackets are used in Table 4), extending the classification introduced by Melo et al. (2005, p. 198), are used to classify the models from literature contained in Table 4:

- *Planning horizon*: Single-period models (S) optimize a network without covering temporal aspects while multi-period models (M) explicitly consider changes expected throughout a planning horizon separated into several time periods.[24] If the periods modeled are medium-term periods such as months or quarters the models are also referred to as multi-season models (cf. Martel 2005, pp. 271-272).
- *Objective function*: Single-period network design models either minimize costs (C) or maximize profits (P) based on accounting data. Multi-period models normally consider the temporal distribution by discounting all values to their present value and are usually based on cash flows instead of accounting data. In the latter case, either present values of expenditures (NEF) or net cash flows (NCF) are optimized but in exceptional cases also the present values of accounting-based costs (PVC) or profits (PVP) might be optimized. Additionally, models with objective functions that optimize multiple objectives (M) simultaneously (e.g., costs and lead times) can be found.
- *Products*: Models are either limited to a single product (S) or capable of covering multiple products (M).
- *Uncertainty*[25]: Uncertainty can be incorporated into strategic analysis in different ways (cf. Kallrath and Maindl 2006, p. 39). From a mathematical point of view, models are either deterministic (D) or stochastic (S). The latter is here defined as explicitly considering a probabilistic repre-

[24] Often, models are characterized as being static or dynamic having the same distinction in mind. This terminology can however be confused with dynamic programming and is therefore not used here.

[25] Some authors further distinguish between decision-making under uncertainty and decision-making under risk with the former referring to a situation where no quantitative information on the "uncertain" parameters is available and the latter referring to a situation where a well-determined probability distribution is available (cf. Wets 1974, pp. 309-310). Here, this distinction will not be used.

sentation of uncertain parameters (cf. Wets 1989, p. 573). Approaches such as scenario- or sensitivity analyses are consequently classified as being deterministic.

- *Geographical scope:* In contrast to purely domestic models (D), international models (I) include factors related to international trade such as tariffs, trade restrictions or currencies.
- *Functional scope:* Both models focusing either on production networks (P) or distribution networks (D) and integrated models (P,D)[26] can be found. Additionally, some models include sourcing decisions (S) such as vendor selection or make or buy decisions and a few models consider marketing aspects (M) for example by modeling competition based on game theory or including pricing decisions.
- *Production levels*: The number of production levels (echelons) can be limited to a fixed number (#) or be unlimited (UL).
- *Distribution levels*: For models that cover production and distribution network, the number of distribution levels (echelons) can similarly be limited to a fixed number (#) or be unlimited(UL).
- *Location levels*: Some models do not allow location choice for all production and/or distribution levels considered. If this is the case, the subset with location choice is specified. If all levels modeled are variable, this is marked, too (all).
- *Capacity*: Both uncapacitated (-) and capacitated (C) models exist. Some of the multi-period, capacitated models also allow expansions (E), relocations (RL) or reductions (R) of capacity throughout the planning horizon.
- *Budget*: Investment budgets are either restricted (R) or unrestricted (-).
- *Inventory*: Inventory is either modeled explicitly (I), not at all (-) or only the effects of network design alternatives on pipeline inventory (PI) are considered by including inventory carrying costs for pipeline inventory.
- *Solution method*: Many models use solution algorithms (A) specifically devised for the optimization problem at hand while others are solved using commercial solvers (C). Additionally, models using heuristics (H) can be found.

---

[26] Production network design by definition needs to include distribution decisions. For the purpose of this classification a model is considered to include distribution network design only if at least one distribution echelon (warehouses, distribution centers, etc.) is explicitly modeled. The term "distribution center" is not defined uniformly in literature (cf. Higginson and Bookbinder 2005). Here, distribution centers where a production step takes place are considered to be a production echelon whereas those performing picking or packaging operations are considered to be a distribution echelon.

- *Application industry*: The majority of the models proposed in literature is of a general nature and not customized to the needs of a specific industry. For industry-specific models the respective industry is noted:
  - *A:* Automotive industry
  - *C:* Chemical industry
  - *E:* Electronics industry
  - *F:* Food processing
  - *P:* Pulp and Paper industry
  - *S:* Steel industry

Table 4. Classification of supply network optimization models

	Planning horizon	Objective function	Products	Uncertainty	Geographical scope	Functions covered	Production levels	Distribution levels	Location levels	Capacity	Budget constraints	Inventory	Solution method	Application industry
Proposed model	**M**	**NCF**	**M**	**D**	**I**	**P**	**UL**	**-**	**all**	**C,E,R**	**-**	**PI**	**C**	**C**
Antunes and Peeters 2001	M	NEF	S	D	D	n/a[3]	1	-	all	C,E,R	R	-	A	
Arntzen et al. 1995	M	M	M	D	I	P,D	UL	UL	all	C	-	I	A	E
Bhutta et al. 2003	M	P	M	D	I	P	1	-	all	C,E,R	-	I	n/a	
Billington and Davis 1992	M[6]	C	M	D	D	P	n/a	-	all	-	-	-	n/a	E
Brown et al. 1987	S	C	M	D	D	P[4]	1	-	all	C	-	-	A	F
Canel and Das 1999	M	P	S	D	D	P	1	-	all	UR	-	-	A	
Canel and Das 2002	M	P	S	D	I	P	1	-	all	C	-	I	n/a	
Canel and Khumawala 1996	M	P	S	D	I	P	1	-	all	C	-	I	C	
Canel and Khumawala 1997	M	P	S	D	I	P	1	-	all	C	-	-	A	
Canel and Khumawala 2001	M	P	S	D	I	P	1	-	all	-	-	-	H	
Canel et al. 2001	M	C	M	D	D	P,D	1	1	D	C,E,R	-	-	A	
Chakravarty 2005	S	P	M	D	I	P,M	1	-	all	C	R	-	A	
Chardaire et al. 1996	M	C	S	D	D	P	1	-	all	-	-	-	H	
Cohen and Kleindorfer 1993[5]	M	P	M	S	I	P	n/a	n/a	n/a	n/a	n/a	n/a	A	n/a
Cohen and Lee 1989	S	P	M	D	I	S,P,D	2	1	-	C	-	-	A	
Cohen and Moon 1991	S	C	M	D	D	S,P	1	-	-	C	-	-	A	
Cohen et al. 1989	M	P	M	D	I	S,P	1	-	all	C	-	-	H	
Dasci and Verter 2001	S	C	M	D	D	P	1	-	all	C	-	-	A	
Dogan and Goetschalckx 1999	M	C	M	D	D	S,P,D	2	-	all	C	-	I	A	
Erlenkotter 1981	M	NEF	S	D	D	P	1	-	all	C,E	-	-	H	
Fleischmann et al. 2006	M	NEF	M	D	I	P	1	-	all	C, E	R	-	C	A
Fong and Srinivasan 1981	M	NEF	S	D	D	P	1	-	all	C,E	-	-	H	
Fong and Srinivasan 1986	M	NEF	S	D	D	P	1	-	all	C,E	-	-	H	
Geoffrion and Graves 1974	S	C	M	D	D	P,D	1	1	D	C	-	-	A	
Geoffrion et al. 1978	S	C	M	D	D	P,D	1	1	D	C	-	-	A	
Gue 2003	M	***	M	D	D	D	-	1	all	C	-	I	C	

Table 4 (continued). Classification of supply network optimization models

	Planning horizon	Objective function	Products	Uncertainty	Geographical scope	Functions covered	Production levels	Distribution levels	Location levels	Capacity	Budget constraints	Inventory	Solution method	Application industry
Proposed model	**M**	**NCF**	**M**	**D**	**I**	**P**	**UL**	**-**	**all**	**C,E,R**	**-**	**PI**	**C**	**C**
Grunow et al. 2006	M	NEF	M	D	I	P	2	-	-	C,E	-	-	C	E
Haug 1985	M	NEF	S	D	I	P	1	-	all	C	-	-	C	
Hindi and Pienkosz 1999	S	C	S	D	D	P	1	-	all	C	-	-	H	
Hinojosa et al. 2000	M	C	M	D	D	P,D	1	1	all	C	-	-	A	
Hodder and Dincer 1986	S	P	S	S	I	P	1	-	all	C	-	-	A	
Hodder and Jucker 1985	S	P	S	S	I	P	1	-	all	-	-	-	A	
Hormozi and Khumawala 1996	M	C	S	D	D	P,D	1	1	D	C	-	-	A	
Huchzermeier and Cohen 1996	M	PVP	S	S	I	S,P	1	-	all	C	-	-	A	
Jacob 2005	M	NEF	M	D	I	P	UL	-	all	C,E,R	R	I	C	
Jayaraman 1998	S	C	M	D	D	P,D	1	1	all	C	-	I	C	
Kalcsics et al. 2000	S	C	M	D	D	P,D	1	1	D	-	-	-	C	
Kallrath 2002[1]	M	M	M	D	D	S,P	UL	-	all	C,E,R	-	I	C	C
Kaufman et al. 1977	S	C	S	D	D	P,D	1	1	all	-	-	-	A	
Kelly and Marucheck 1984	M	NEF	S	D	D	D	-	1	all	C	-	-	A	
Kendrick and Stoutjesdijk 1978	M	NEF	M	D	D	P	UL	-	all	C,E,R	-	-	n/a	
Klincewicz and Luss 1987	S	C	M	D	D	P	1	-	all	-	-	-	A	
Klincewicz et al. 1988	M	NEF	S	D	D	D	-	1	all	C,E,R	-	-	H	
Köksalan and Süral 1999	M	NEF	M	D	D	P	1	-	all	C	-	-	n/a	F
Kouvelis and Rosenblatt 2002	M	NCF[6]	M	D	I	P	2	-	all	C	-	-	C	
Kouvelis et al. 2004	M	NCF[6]	M	D	I	P	2	-	all	C	R	-	C	
Lee and Luss 1987	M	C	S	D	D	P	1	-	-	C,E,RL	-	-	A	
Martel 2005	M	P	M	D	I	S,P,D,M	UL	UL	all	C,E,R	-	I	H	
Martel et al. 2005	S	P	M	D	I	S,P,D	2	1	all	C	-/R	I	C	P
Maßmann 2005	M	P	S	S	D	P	1	-	all	C	-	-	A	
Mazzola and Neebe 1999	S	C	M	D	D	P	1	-	all	C	-	-	A,H	
Melachrinoudis and Min 2000	M	M	S	D	D	P	1	-	all	C,E,R,RL	R	-	C	
Melo et al. 2005	M	C	M	D	D	P,D	1	UL	all	C, E,R,RL	R	I	C	
Meyer 2005	S	C	M	D	I	P	UL	-	all	C	-	PI	C	
Mohamed 1999	M	C	M	D	I	P	1	-	-	C,E,R	-	I	n/a	
Nickel et al. 2005	M	C	M	D	D	S,P,D	1	UL	all	C,E,R,RL	R	I	C	S
Papageorgiou et al. 2001	M	NCF	M	D	I	P	1	-	all	C,E	-	I	C	C
Paquet et al. 2004	S	C	M	D	D	S, P	UL	-	all	C	-	-	A	
Pirkul and Jayaraman (1996)	S	C	M	D	D	P,D	1	1	all	C	-	-	H	
Pirkul and Jayaraman (1998)	S	C	M	D	D	P,D	1	1	all	C	-	-	H	
Pomper 1976	M	NCF	S	D	I	P	1	-	all	C,E	-	-	H	
Pooley 1994	S	C	M	D	I	P,D	1	1	all	C	-	-	C	F
Sabri and Beamon 2000	S	C	M		D	S,P,D	1	1	all	C	-	-	C	
Sankaran and Raghavan 1997	S	C	S	D	D	P	1	-	all	C	-	-	C	

Table 4 (continued). Classification of supply network optimization models

	Planning horizon	Objective function	Products	Uncertainty	Geographical scope	Functions covered	Production levels	Distribution levels	Location levels	Capacity	Budget constraints	Inventory	Solution method	Application industry
Proposed model	**M**	**NCF**	**M**	**D**	**I**	**P**	**UL**	**-**	**all**	**C,E,R**	**-**	**PI**	**C**	**C**
Schmidt and Wilhelm 2000	S	P	M	D	I	P	UL	1	all	C	-	-	n/a	
Shulman 1991	M	C	S	D	D	P	1	-	all	C,E	-	-	A	
Syam 2000	M	C	S	D	I	P	1	-	all	C,E	-	-	H	
Tcha and Lee 1984	S	P	S	D	D	D	-	UL	all	-	-	-	A	
Tragantalerngsak et al. 2000	S	C	S	D	D	D	-	2	all	C	-	-	A	
Truong and Azadivar 2005[2]	S	C	S	D	D	P,D	UL	UL	all	C	-	PI	C	
Tsiakis et al. 2001	S	C	M	D	D	P,D	1	2	D	C	-	-	C	
Van Roy and Erlenkotter 1982	M	NEF	S	D	D	P	1	-	all	-	-	-	A	
Verter and Dincer 1995	S	C	S	D	D	P	1	-	all	C	-	-	A	
Vidal and Goetschalckx 1996	S	C	M	D	I	S,P	1	-	all	C	-	PI	C	
Vidal and Goetschalckx 2000	S	C	M	D	I	P	1	-	all	C	-	PI	C	
Vidal and Goetschalckx 2001	S	P	M	D	I	P,D	1	1	-	C	-		H	
Wouda et al. 2002	S	C	M	D	D	P	2	-	all	C	-	-	C	F

[1] Combined strategic and operational model
[2] Genetic algorithm and simulation model to analyze stochastic effects
[3] Public services application
[4] Includes allocation of production facilities (equipments) to plants
[5] No mathematical formulation provided
[6] Investment decisions only allowed in first period of planning horizon

As discussed in Chapter 2.1.2, this work focuses on production network design. Consequently, the classification only lists models that include production network design decisions. For reviews and further references related to distribution network design the reader is referred to Tsiakis et al. (2001), Geoffrion and Powers (1995) or Aikens (1985).

3.2.2 Review of Individual Publications

Application-oriented Publications

Grunow et al. (2006) present a model developed for supply network design in electrical components manufacturing. The model focuses on product relocation and capacity expansion decisions and does not consider setup of new plants or closure of existing ones. Additionally, simple assembly plants supplied with complete kits from a major production site are modeled. To account for production-development synergies in early life-cycle stages of a product, relocation of relatively new products can be restricted. Furthermore, personnel capacity restrictions for the experts required to support relocations and capacity expansions are included. A combination of fixed cost charges for every product family allocation and restrictions on the maximum number of allocations is employed to capture the effects of product mix complexity. The model considers tariffs and local content rules in simplified form but no other aspects of international trade such as currency effects.

Fleischmann et al. (2006) provide a global production network planning model used at BMW that extends the simpler load planning model proposed by Henrich (2002). The model is a multi-period, multi-product model with an objective function that maximizes the pre-tax net present value of the network. It includes decisions on product-plant allocation, production volumes, material sourcing volumes by supply region, structural and product-specific investments and use of overtime capacity. A major contribution of the model is the incorporation of the time-distribution of investment expenditures typically observed in automobile production networks. While tariffs are included in the transportation costs, the model does not consider further aspects of international trade such as currencies, duty drawbacks or local content rules which play a major role in practice.

Jacob (2005) proposes a model that seeks to cover a broad range of discrete parts production systems. The model maximizes the net present value of a multiple-echelon, multiple period, multiple product global production network. Features include the explicit modeling of reinvestments required after the expected economic life of assets has been reached, a comprehensive modeling of personnel effects (e.g., training needs for newly hired employees, personnel costs for different skill levels, costs for expatriates and management resources available for network optimization projects) and ramp-up costs associated with establishing new production sites. Productivity improvements are integrated via annual productivity growth rates that, contrary to common assumptions, are not based on cumulative production volumes. With respect to international trade, government subsidies and tariffs (but not duty drawbacks) are included in the model. However, the complete model formulation is not provided. A separate model is proposed to determine the best production technology for each process-site combination prior to specifying the network design model. Four application cases ranging from automotive components to electronics industry are reported.

Martel et al. (2005) propose a single-period, deterministic, multiple-product MILP model specifically developed for optimization of global pulp and paper supply chains. Based on the characteristics of pulp and paper supply chains two production echelons, one distribution echelon and location-specific raw material availability are considered. Capacity can be selected from a set of discrete alternatives and dedicated and flexible technologies are available. The features of international trade modeled are currencies, cross-border trade costs, local content rules and taxes. Extensions to treat transfer prices as a variable to reduce the tax burden and to explicitly include financing options are also proposed and approaches to consider exchange rate uncertainty are discussed.

Martel (2005) proposes a comprehensive model for the design of global production-distribution networks covering aspects such as capacity modifications, technology selection, vendor selection, etc. taking a medium-term, multi-season perspective. Special features are the modeling of alternative marketing policies and the comprehensive consideration of inventory in the form of order cycle stock, safety stock and seasonal stock including risk pooling effects. Marketing policy options (e.g., alternative delivery times) are included by allowing to choose between discrete site-market-marketing policies. The policy selected via a binary variable determines the price that can be realized for a product. The chosen inventory modeling approach leads to a non-linear relationship between inventory level and throughput and thus the model is a MINLP model.

Specific aspects of dairy production are covered by the model proposed by Wouda et al. (2002). While the model is relatively simple with respect to its mathematical structure (single-period, domestic setting, two production levels), it contains aspects such as milk collection from production regions, divergent material flows occurring in the first stage of dairy processing and transshipment of the respective intermediates between plants.

Valuable insights into how to model aspects of chemical production networks can be found in the model Papageorgiou et al. (2001) propose for use in the pharmaceutical industry. The model is a multi-period, multi-product global optimization model that also includes decisions on (candidate) product portfolio management. The objective function maximizes the net present value of after-tax cash flows. Plants are separated into "blocks" containing production "suites" and service centers that provide utilities, etc. A fixed relationship between the number of suites and the number of service centers required to support them is assumed. All suites are defined to possess identical capacities but production rates are suite- and site-specific. A major contribution of the model is the way it accounts for specifics of pharmaceutical active ingredient production. The items covered include qualification requirements, scale-up processes and the capacity requirements for changeover processes which are typically very complex in pharmaceutical production. Additionally, the taxation aspects of a trading structure involving an intellectual property center (principal trading company) are covered. Since inventory buildup is permitted, product lifetime constraints are included, too. Deficiencies of the model are that it does not cover duties, exchange rates and multi-level production systems. Also, the product portfolio selection approach appears to be overly simplistic considering the complexities of pharmaceutical product portfolio management.

One of the most comprehensive production network design models published so far is the model from Digital Equipment Corporation as reported by Arntzen et al. (1995). The model is a multi-period, multi-product model covering several production levels and various aspects of international trade such as duties and duty drawbacks and local content requirements. In addition to location selection also facility configuration is optimized. The objective function takes the form of a multiple-criteria optimization model containing both cost items and lead time aspects which are combined using a weight factor. While the authors report the successful application of the model at Digital Equipment no solution methodology is provided and no numerical results are reported.

Recently Published Models

Chakravarty (2005) integrates overhead allocation and pricing decisions into a single-period, deterministic, multiple-product plant location model. The location model itself is relatively simple. Only one production stage can be modeled, capacity is assumed to be a continuous variable without upper or lower bounds and investment expenditures are linearly related to capacity. The complexity of the non-linear model is caused by including pricing and overhead allocation decisions. Demand is described by price-dependent demand curves in a non-competitive environment. The amount of overhead absorbed by each product is also determined by the model. Extensions to the basic model integrate fixed investment expenditures, exchange rates, taxes, local content rules and limits on market volumes. While being of limited applicability to real-world production networks, the model can be used to analyze the effects of tariffs and local taxes on optimal overhead allocation in international production networks.

Melo et al. (2005) propose a multi-period, deterministic, multiple-product MILP model for strategic supply chain planning. The model does not impose any restrictions on the number and type of facilities and the transportation links between facilities. The basic model explicitly covers relocation of capacity to new facilities. It can be extended to include capacity expansions and reductions. To this end, two fictitious, non-selectable facilities are introduced that provide additional or absorb excessive capacities. Capacity is assumed to be adjustable on a continuous scale but an extension to modular capacity is also provided. The model is very versatile. However, while Nickel et al. (2005) provide an application example from steel industry, the underlying assumptions limit the model's general applicability to problems from industry. Only a single production stage can be modeled. Capital expenditures are not included in the objective function but instead only restricted by a budget limit. If the budget is not fully used in one period, the remainder including interest earned is available in subsequent periods. This approach does not ensure that the savings in operating costs are sufficient to justify the investments required to achieve them. Additionally, fixed facility costs are independent of capacity even in the model extensions allowing changes to the overall network capacity. While expenditures for capacity changes are allocated to the time period preceding or following the implementation to reflect the time-consuming nature of these activities, the fact that capacity relocations effectively render the capacity unusable for the time of the transfer is not considered. Instead it is assumed that the duration of the relocation is relatively short compared to the time period (cf. Melo et al. 2005, p. 185).

Truong and Azadivar (2005) propose what they refer to as a "hybrid optimization approach" to integrate make or buy decisions, supplier selection, production planning policy, transportation mode selection, facility location, capacity, production and service allocation into a model minimizing total system costs. Their model consists of three components. A genetic algorithm is applied to determine make or buy choices, select suppliers, transportation modes and production policies. A MILP calculates optimal location, capacity, production and service allocation decisions based on these structural decisions. The results are evaluated via a simulation tool and the process is iterated until stopping criteria are met. Decision scope and evaluation criteria available are comprehensive. However, the model is a single-period model, limited to a single product and a domestic setting, restricting its applicability to real-world production systems.

Fandel and Stammen (2004) integrate product development and product recycling aspects into an extended global supply chain model that covers procurement, several production levels and two distribution levels. Their model optimizes the after-tax net present value of the supply chain with decision variables including the selection of research, production, distribution and recycling facilities, product flow and storage quantities and product development decisions. An interesting feature of the model is the use of macro (years) and micro (quarters/months) time periods which was derived from lot sizing and scheduling models. Development and financial parameters are determined on the level of macro periods while material flows, sales volumes and revenues are determined at the micro level. All network design decisions are however fixed at the beginning of the planning horizon. The model can support multiple production levels but all variable unit costs have to be developed outside the model, capacities cannot be selected and no production technology selection is possible. Hence, its applicability to real-life production systems seems questionable even though the authors claim that automotive industry might be a suitable industry. The resulting model is very complex due to the extended scope and the authors report that only small problem instances have been solved so far and no experiences from real-world applications are provided.

Paquet et al. (2004) propose a single-period, domestic production network design model that includes technology selection (flexible/dedicated). Besides this aspect the main features of their model are that the network structure is not restricted (i.e., also allowing transshipments between plants), vendor selection is included and complex product structures can be accommodated. Comparing a solution algorithm based on Bender's decomposition with the use of CPLEX 6.6, the authors find that in many cases commercial solvers are capable of finding a solution within shorter computation times. While the model covers many features required for application in industry its main drawbacks are the single-period formulation and the focus on a domestic setting.

Kouvelis et al. (2004) present a relatively simple multi-period MILP plant location model for global production network design with investment decisions only allowed in the first period. The production system consists of component-dedicated manufacturing sites and final assembly sites. It is limited to two production levels and one final product. The objective function maximizes the NPV of the production network. The main purpose of the model is to analyze the effects financing subsidies, tax regimes, tariff structures and local content requirements have on optimal network design. The analysis is based on theoretical considerations and a numerical example. More complex aspects of international trade such as duty drawbacks are not considered.

The model proposed by Bhutta et al. (2003) is a multi-period, deterministic multiple-product MILP model integrating plant location, production, distribution and investment planning in a global environment. It is relatively simple both mathematically (no binary decision variables but integer production quantities) and with respect to the assumptions made for key modeling parameters. Capacity can be modified continuously without lower or upper bounds. International features are limited to exchange rates and tariffs.

3.3 Modeling Specialty Chemicals Production Networks

Based on a comparison of alternative proposals from literature, in this chapter modeling approaches are developed that cover the specific requirements of specialty chemicals production networks. The focus is not so much on mathematical aspects but on applicability to problems from industry.

3.3.1 General Model Characteristics

Table 4 in Chapter 3.2.1 already characterizes the model developed in this work but does not provide the underlying rationale. Below, this rationale is given for all classification criteria except the choice of objective function to which Chapter 3.3.2 is dedicated.

Planning Horizon

Different opinions exist with respect to the type of planning horizon appropriate for strategic analyses. For example, Hanssmann (1990, pp. 324-328) argues that a single-period planning horizon is sufficient in strategic planning. Contrarily, Freidenfelds (1981, p. 50) postulates that at least from a conceptual perspective an infinite horizon should be used. The specialty chemicals industry is a mature industry where companies already operate extensive production networks. To support strategic planning, a production network design model has to provide insight into how to adapt such an existing network to changes in the business environment and thus a multi-period model is required (cf. Goetschalckx and Fleischmann 2005, p. 131; Henrich 2002, p. 182). Furthermore, single-period models face the risk of ignoring the impact of cost changes such as quickly rising wage levels (cf. Bartmess and Cerny 1993, pp. 86-87). Typically, planning horizons used are in the range of 5-10 years (cf. Billington and Davis 1992, p. 590; Pooley 1994, p. 116; Olbert 1976, p. 116) with each period lasting one year. In the case study described in Chapter 5 a 10 year planning horizon is used.

Products

Due to the fact that multiple products with different processing requirements share production equipment and internally produced intermediates have to be included in the analysis, a multiple-product model is required.

Uncertainty

As pointed out above the choice to be made is whether to employ a stochastic programming model or to integrate uncertainty inherent in the parameters influencing the network design decision via approaches such as scenario analysis. The latter approach is chosen here for several reasons. Firstly, stochastic models are more difficult to solve to optimality (cf. Vidal and Goetschalckx 2000, p. 98). Secondly, developing meaningful probability distributions for uncertain parameters in strategic models is in itself an extremely difficult task (cf. Popp 1983, p. 45). Strategic decisions occur only infrequently and parameters have to be estimated over a long planning horizon rendering infeasible commonly used approaches such as employing historical data or building on previously gathered experiences. Additionally, due to the multi-period nature of the model, transit probabilities would be required to link the parameter realizations across time periods. Thirdly, the concepts underlying stochastic optimization are difficult to communicate to decision makers in top management. Finally, a robust optimization approach based on a limited number of discrete scenarios can be chosen if uncertainty is to be explicitly considered. In contrast to stochastic optimization, robust optimization seeks to reduce the variance of the solution and can account for risk aversion of decision makers (cf. Mulvey et al. 1995, pp. 269-270). Robust optimization is further discussed in Chapter 3.4.4. For an introduction to stochastic programming the reader is referred to Wagner (1969, pp. 638-686) and for stochastic location models to Maßmann (2005).

Geographical Scope

As discussed in Chapter 2.3.4 specialty chemicals companies operate global production networks. Additionally, tariffs often constitute a higher share of total costs than transportation costs. Thus, a model that explicitly considers international trade is required.

Functional Scope

As argued in Chapter 2.1.2 this work focuses on production network design. Demand is assumed to be independent of production network design decisions and the cost structure achieved. Also, unlimited quantities of raw materials are assumed to be available. An extensions to include make-or-buy decisions (and possibly vendor selection) is provided to accommodate application cases where sourcing decisions have to be included.

Production Levels

The number of production levels that have to be modeled explicitly depends on the type of value chain considered and the level of vertical integration pursued by the company. Consequently, the model does not structurally limit the number of levels. However, from a computational point of view limits might still apply.

Capacity

Clearly, uncapacitated network design models are of limited use in practice. Instead, models need to be capable of expanding and reducing capacity to reflect the dynamic nature of production network design and the fact that an existing network is the starting point for the analysis. However, due to the tight integration of chemical production equipment into plant buildings a physical transfer of capacity from one plant to another (relocation of assets) is an extreme exception that does not have to be considered.

Budget Constraints

In theory, a company should be able to finance any investment that earns a return exceeding the cost of capital and hence no budget constraints would be required. Additionally, Melo et al. (2005, p. 201) observed that budget restrictions can lead to increased calculation time. Primarily for the first reason the proposed model does not contain a restriction on the investment budget. If required a budget restriction can easily be added to reflect the fact that in practice investment budgets are often limited.

Inventory

Strategic planning usually is based on yearly periods. Hence, inventory levels only have to be modeled explicitly if capacity requirements fluctuate between years and peaks can be leveled by building up inventory and carrying this inventory over from year to year. Combinations of product life cycles, technology dedication and durability of goods generating such a demand pattern are an exception in specialty chemicals and thus here production volumes are set to equal sales volumes. Papageorgiou et al. (2001, p. 280), who model inventory considering product lifetime constraints for pharmaceutical industry, do so primarily because "companies find it useful to be able to plan year-end inventories".

Furthermore, inventory carrying costs arising from seasonality of demand within a time period are not considered either. Following Chakravarty (2005, p. 162), a capacity cushion to balance the costs of capacity shortages and inventory carrying costs can be determined outside the network design model (cf. Hayes and Wheelwright (1984, pp. 48-54) for details on determining capacity cushions). Alternatively, the linear interpolation proposed by Goetschalckx and Fleischmann (2005, p. 132) could be used. However, since production network design has a significant effect on product volumes tied up in the supply chain, pipeline inventory has to be considered.[27]

Solution Procedure

The proposed model is solved using commercial optimization software. Algorithms designed to exploit characteristics of a specific problem tend to have shorter solution times but require substantial mathematical expertise and make modifications difficult thus limiting their applicability in industry. Heuristic solution procedures, in addition to facing the same problems as specific solution algorithms, often make comparisons of results obtained in different runs of the model impossible (cf. Geoffrion and Powers 1980, pp. 24-25). As inter-run comparisons are a critical element of network design analyses, heuristic approaches should be avoided in this context.

3.3.2 Objective Function

Choice of Objective Function

As can be seen in the model classification in Chapter 3.3.1 different types of objective functions are used in production network design models. Goetschalckx and Fleischmann (2005, p. 121) point out that the objective of employing strategic network design models is to identify network design alternatives that maximize the long-term economic performance of the corporation. In order to do so, the objective function has to maximize the contribution of the production network to the overall value of the company. A range of methods that differ with respect to complexity and underlying assumptions are available to quantitatively evaluate investment alternatives.[28] In a multi-period model that includes decisions on the temporal allocation of investments a dynamic evaluation method is required. Within the sub-segment of dynamic investment appraisal methods, as Brealey et al. (2006, pp. 84-111) argue, only the net present value (NPV) method leads to investment appraisals aligned with the objective of maximizing a

---

[27] The reader interested in a detailed discussion of inventory management is referred to Tempelmeier (2006) and to Graves and Willems (2003) for a discussion of how to spread safety stocks across the supply chain.

[28] For an overview of investment appraisal calculation methods see for example Götze and Bloech (2002) or Perridon and Steiner (2006).

company's value. Similarly, in the literature on capacity planning the argument for using the NPV as decision criterion has been made (cf. Freidenfelds 1981, pp. 10-70). As the NPV method is cash-flow based, the model structure can at the same time be used to study the effects of exchange rate fluctuations which is an important issue in global supply networks. (cf. Cohen and Kleindorfer 1993, p. 16). It should be noted however that other "order winning criteria" (cf. Hill 1992, p. 10) such as delivery lead times, quality, etc., which cannot be easily integrated into the objective function of a strategic optimization model, may be affected by the selected production network design as well.

The principle of the NPV method is to forecast over time all cash flows associated with an investment. Each period's net cash flows are then discounted to the present.[29] As discount rate usually the company's cost of capital is used because in this case a positive NPV indicates that the investment increases the company's value (cf. Rappaport 1998, p. 37; see Appendix 1 for a detailed discussion of how to derive the appropriate discount rate). The calculation of the NPV is based on the following formula:

$$NPV = -I_0 + \sum_{t=1}^{N} \frac{C_t}{(1+r_r)^t} + \frac{TV_N}{(1+r_t)^N} \tag{3.1}$$

where: I_0 = Initial investment
C_t = Net cash flow at the end of period t
TV_N = Terminal value at the end of period N
r_t = Discount rate in period t
N = Planning horizon

The objective function of a supply network design model can either minimize costs or maximize profits. In practice the production function is often required to assume that all forecasted demands have to be met. In this constellation cost minimization and profit maximization lead to identical results and consequently cost minimization models are used. From an economic perspective this simplification can be justified in cases where a high share of fixed costs allows the assumption that any product sale con-

[29] Assumptions on temporal allocation of cash flows and discounting period need to be consistent (e.g., beginning, middle or end of period) which is sometimes not the case in mathematical optimization models (cf. Erlenkotter 1981, p. 134). Generally, it is assumed that cash flows are realized at the end of a period. Alternatively, continuous payments can be assumed but the error caused by the year-end assumption is limited (cf. Brealey et al. 2006, pp. 46-48).

tributes towards covering fixed costs. If this assumption does not hold, it is possible that the model proposes decisions which actually reduce the company's value. As variable costs often constitute between 75 and 85 per cent of total production costs in specialty chemicals (cf. Chap. 5.2.3), the full potential of a network design model is only realized if the model can also decide whether or not to serve certain demands. Hence, an objective function that maximizes the net present value of cash flows is required.

End-of-Horizon Effect

The use of a cash flow-based objective function in multi-period optimization models gives rise to a structural problem. As shown above, the NPV method is based on cash flows forecasted for a limited planning horizon and a terminal value to account for time periods not explicitly modeled. In practice, the terminal value is often assumed to be zero when assessing individual investments because typically companies not only require a positive NPV but also a dynamic payback period that is shorter than the planning horizon. However, in network design models investment decisions can also be taken at the end of the planning horizon. In this case, the remaining time periods of the planning horizon might not be sufficient to recapture the investment expenditure (cf. Fig. 17). At the same time it is not possible to estimate the terminal value of such investments because the investments themselves are not known up front. As a consequence, NPV-based models in certain constellations show an *end-of-horizon effect* whereby investment opportunities at the end of the planning horizon cannot be properly evaluated. This problem does not occur with accounting-based objective functions because here investments are capitalized and only the annual depreciation is considered.[30]

In the recent literature, the end-of-horizon effect has not been adequately discussed. Some of the NPV-based models reviewed, mostly without explicitly saying so, introduce additional assumptions that reduce the likelihood of occurrence or entirely eliminate the end-of-horizon effect. For example, Kouvelis et al. (2004) and Kouvelis and Rosenblatt (2002) assume that investment decisions have to be taken at the beginning of the planning horizon and plants remain open. Melo et al. (2005) and Antunes and Peeters (2001) do not include investment expenditures in the objective function but solely restrict them via budget constraints. Canel and Khumawala (1997) maximize the NPV of cash flows, explicitly stating that

[30] Note that in case of restructuring costs that cannot be capitalized models with an objective function based on accounting values face the same problem.

terminal values are not considered. The end-of-horizon problem does not occur because all demands have to be met and capacity reductions are not possible.

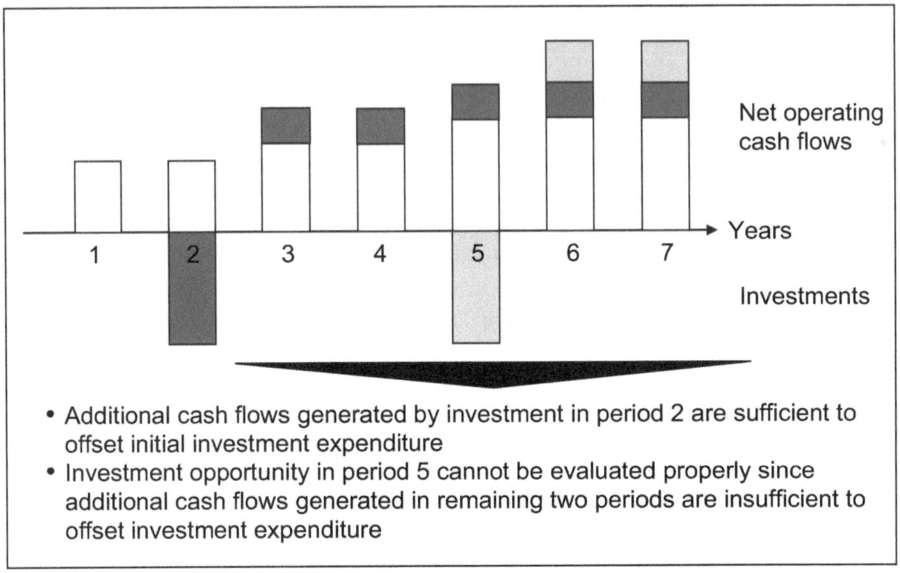

Fig. 17. End-of-horizon effect

In other cases the problem is not discussed even though it can occur. For example, the model proposed by Papageorgiou et al. (2001) uses an objective function maximizing NPV without requiring that all demands are met. In the numerical example provided, total demand in period 5 of the 10 periods planning horizon already exceeds 95% of demand in the final period and capacity reductions are not modeled. Hence, the problem most likely does not occur in the specific data instance the publication is based on. Similarly, Haug (1985), while maximizing NPV and Roodman and Schwarz (1975), who model facility phase-out strategies including closure costs, do not discuss the effect at all. Likewise, many long-range capacity planning models for an individual site that are based on an NPV objective function ignore the end-of horizon effect (cf. Bok et al. 1998; Sahinidis et al. 1989, pp. 1050-1051).

The end-of-horizon problem can be resolved in different ways. The simplest option is to accept that network design models with a cash flow objective function cannot properly evaluate investment and restructuring decisions at the end of the planning horizon if their dynamic payback exceeds the remaining periods of the planning horizon. Alternative approaches can be found in literature from the 1960ies and 70ies where the end-of-horizon effect is discussed explicitly. Kendrick and Stoutjesdijk (1978, pp. 39-43) and Kendrick (1967, pp. 21-22) suggest to convert the capital expenditures of investment projects into equivalent uniform payment series and cut these payments off at the end of the planning horizon. Thus, only the proportion of the investment expenditures that is actually "utilized" during the planning horizon has to be recaptured, effectively simulating a leasing/rental process.

Pomper (1976, p. 150) explicitly estimated terminal values and integrated them into his model. He calculated the additional cash flows generated by each alternative network configuration in the final period of the planning horizon as compared to the "do-nothing" option. To determine terminal values incremental cash flows were capitalized over a period of 10 years. However, this approach was only possible because all alternative final-period network configurations were known up front due to the backtracking solution procedure employed.

Without reference to Pomper, Eppen et al. (1989, p. 520) implemented a similar logic in a multi-period capacity planning model for the automotive industry by assuming that external parameters remain constant after the final period of the planning horizon. The last period is extended to infinity in order to eliminate the end-of-horizon effect with no structural decisions being allowed to take place in this terminal "period". This principle can also be employed to allow a model to estimate terminal values without extending the horizon to infinity. By replicating the data from the final period of the planning horizon for the number of periods reflecting the maximum dynamic payback accepted by the company and fixing all structural variables to the values realized in the last period of the planning horizon, the model can use these cash flows to evaluate decision alternatives available in the final periods of the planning horizon.

Business valuation literature provides various other methods for estimating terminal values (for an overview see Koller et al. 2005, pp. 271-290). Unfortunately, as cash flows cannot be allocated to individual decisions in a network design model, a cash flow-based estimate is not possible. Instead, book value or liquidation value at the end of the planning horizon could be used. For example, Fong and Srinivasan (1981, p. 790) include a terminal value function in the unit capacity acquisition cost function. However, they do not specify how this function can be quantified in real-world applications. The major disadvantages are that it is difficult to justify the assumptions underling the terminal value estimate and that restructuring expenditures cannot be properly evaluated.

As investment decisions are in practice only taken for the first periods of the planning horizon, a feasible option is to split the planning horizon into two parts and allow capacity decisions to occur only in the first part of three to five years. This would not impede the value of the network design model as a decision support tool. Instead, the model is re-applied regularly in a rolling horizon approach with updated data. However, if this approach is chosen it has to be considered when creating the data forecast series because volume/price trends extending beyond the first part of the planning horizon where capacity adjustments are allowed could lead to distorted capacity decisions, too.

3.3.3 Capacity Selection, Expansion and Reduction

Technical Capacity

Different approaches to modeling technical capacity have been proposed both in network design literature and in the literature focusing on capacity planning within existing networks.[31] Classically, models assume that capacity is a continuous variable with linear capacity acquisition costs (cf. Truong and Azadivar 2005, p. 2225; Verter and Dincer 1995, p. 1143). Verter (2002, p. 584) and Dasci and Verter (2001, p. 964) also employ a continuous capacity variable but consider technology alternatives that lead to scale economies in capacity acquisition. The result is a monotone increasing concave capacity acquisition cost function. The capacity planning model proposed by Sahinidis et al. (1989, p. 1050) for long-range planning in chemical industry combines linear capacity selection with a fixed-cost charge incurred at each expansion to account for economies of scale in capacity installation. Bhutta et al. (2003, p. 204) allow both continuous capacity selection/expansion and capacity reduction. Melo et al. (2005, p. 189) propose a continuous capacity variable in a model that allows to select, expand and reduce capacity and explicitly models transfer of capacity between old and new plants. Additionally, an extension to modular capacity is presented. Nickel et al. (2005) report an application of the basic, continuous version of this model in steel industry.

An alternative to the assumption of continuous capacity is to model capacity discretely. Lee (1991, p. 169) and Haug (1985, p. 92) propose models which select for each potential site the plant to be constructed from a set of possible plant types with pre-defined capacities. Mazzola and Neebe (1999, p. 286) also use this approach and Sankaran and Raghavan (1997,

[31] As Luss (1982, p. 935) points out, capacity planning in multi-facility networks is closely related to network design problems.

p. 4) present an application example for locating and sizing propane bottling plants. Shulman (1991, p. 425) extends the modular capacity concept to allow for discrete capacity expansions in a multi-period setting. His model can establish multiple facilities at a given site. One variant only allows to combine identically sized facilities (one facility type) at a given site while another variant also allows to combine several facility types. Similarly, Syam (2000, p. 176) uses three discrete capacity sizes from which to select initial capacity and capacity expansions in later time periods (restricted to one per period). Expansion costs are modeled as a stepwise decreasing function leading to piecewise linear expansion costs.

As discussed in Chapter 2.3.3, the technical capacity of a specialty chemicals plant is determined by the size of individual production lines and the number of parallel installations. Theoretically, a production line's technical capacity can, within certain bounds, be selected on a continuous scale. For modeling purposes it is however sufficient to restrict this choice to a few discrete size classes (e.g., small, medium and large) because production equipment is often available in certain standard sizes and creating cost-functions for a continuous model is practically impossible. In practice it is additionally preferred to operate only similarly sized production lines at a given plant and this assumption can also be found in publications dealing with batch process capacity planning (e.g., Petkov and Maranas 1998, p. 790). The underlying reason is that process characteristics such as heat exchange are influenced by equipment size. Hence, operating different equipment size classes would either make a product transfer between production lines impossible and thus reduce flexibility or require the development of a specific recipe for each production line. Additionally, in cases where shared resources exist (cf. also Chap. 3.4.3) production line-dependent batch sizes would further complicate production planning. Consequently, a discrete, modular approach with the possibility of extending or reducing the number of production lines over time, assuming identically sized production lines for a given plant, is proposed to model technical capacity.

Unit capacity acquisition costs in chemical production equipment typically show strong economies of scale. While the degree of scale economies varies between plant types a rough estimate used in industry is the assumption that doubling capacity leads to acquisition cost increase of 50%. Ignoring the interaction with capacity mix effects, the nature of the capacity acquisition cost function favors larger and fewer production lines and would in some constellations lead to the construction of plants with a single production line. Since this would significantly reduce the flexibility of the plant and, in the case of a single production line, also create an over-dependency on the equipment, a restriction defining a minimum (and maximum) number of parallel production lines is included in the model. Options to explicitly consider product mix effects are discussed below in Chapter 3.3.4.

In addition to the choice of capacity, firms can often choose between alternative production technologies that may have an effect on equipment flexibility, service times, quality or reliability as well as costs. Commonly, if technology selection is considered, it is in the context of product-dedicated versus flexible equipment (cf. Paquet et al. 2004, pp. 118-120; Verter and Dincer 1992, pp. 13-15). Technology choices may also exist with respect to the resulting variable unit manufacturing costs. For example, Verter (2002, pp. 584-585) links alternative technologies with different monotone increasing concave unit manufacturing cost functions. However, in chemical industry technology choice is largely determined by requirements of the underlying chemical processes.

Nevertheless, variable production costs can be influenced via the degree of automation. The main effect of increasing the level of automation is a reduction of the number of operators required to run a plant. Therefore, in addition to selecting the optimal size class, the model has to be able to choose between different degrees of automation. Additional effects of increased automation (e.g., higher process stability) exist, but the economic impact of these effects can often not be properly quantified and will thus not be considered here. Again, for purposes of simplification it is proposed to use discrete levels of automation and to assume an identical level of automation within a plant. The latter assumption, while realistic for newly established plants, might be violated at existing plants. If this is the case an average level of automation should be assumed to avoid having to model a dependency between unit production costs and the production line the product is produced on. Considering this level of detail would clearly be beyond the requirements of strategic planning. Similarly, the technical design of production equipment with respect to the capacity of individual units and their configuration is not considered. The reader interested in this subject can find references on plant design for example in Yang (2004, pp. 50-51).

The combinations of size classes and automation levels (cf. Fig. 18 for an example) describe the set of plants from which the model can choose when setting up new plants. As shown in the example not all theoretical combinations of capacity classes and degrees of automation might be feasible and additional, location-specific restrictions might apply. For expansions of existing plants the choice is restricted to adding production lines that fit into the production environment already in place.

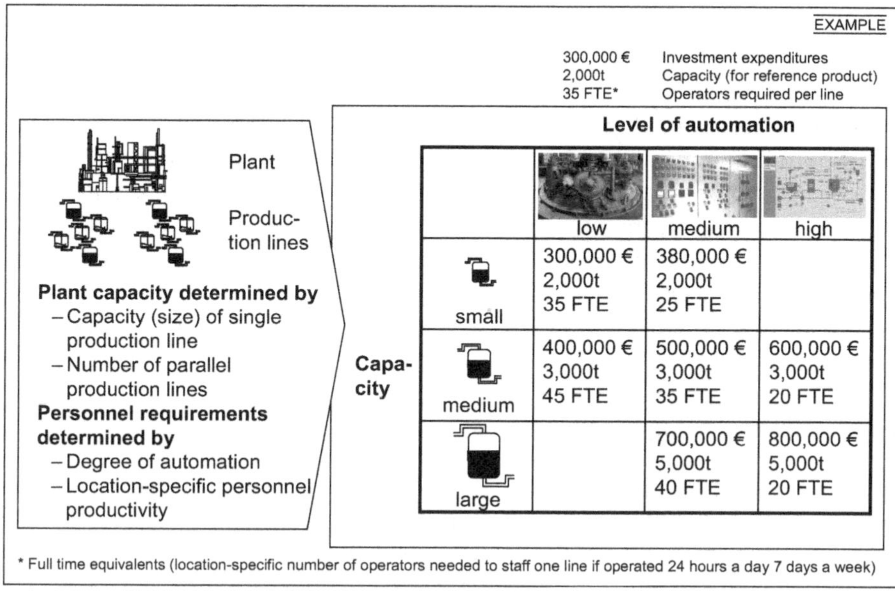

Fig. 18. Example of plant matrix for a single plant class

Personnel Capacity

Personnel effects are often not considered explicitly in production network design models. Haug (1985, pp. 89-90) determines the number of workers that have to be hired per time period based on attrition rates, learning curve effects and volume increases. He includes training costs for newly hired workers (adjusted for government-provided training grants). Also, a maximum number of employees can be defined for each site. Jacob (2005, pp. 82-86) extends this approach by including severance payments associated with personnel reductions and by explicitly modeling costs for expatriates required to set up and operate foreign production facilities.

For specialty chemicals significant hiring only occurs in case of technical capacity increases. Hence, training costs can be included in capacity acquisition expenditures. Since severance payments often dominate restructuring costs (cf. Jacob 2005, p. 54) and the social implications and implementation challenges associated with strong or frequent changes in personnel requirements are important factors, personnel capacity has to be modeled explicitly and the extent of changes allowed per time period may have to be restricted. The proposed model ties fixed personnel to structural decisions and variable personnel to production volumes. While the distribution between fixed and variable personnel may be an object of intense discussion in practice, at the strategic level it is usually possible to align staffing with production levels via shift models, plant holidays, production line closure, etc.

Another element of personnel capacity can also affect production network optimization. Each capacity modification and product transfer project requires the expertise of the process development or technical engineering staff. Consequently, the capacity available with these departments can restrict the number of projects that may be pursued in any given time period. Grunow et al. (2006, pp. 6-8) integrate a restriction into their model to incorporate this aspect. While the basic model proposed in this work does not contain such a restriction, it could be easily extended accordingly.

Capacity of Utilities

In literature, utility capacity is rarely treated. In a simplified approach, Papageorgiou et al. (2001, p. 277) require that a maximum of four "production suites" can be attached to a "service center" providing all utilities required. As discussed in Chapter 2.3.2, utilities such as wastewater treatment, heating or cooling processes constitute a considerable share of a chemical site's cost base. Due to other production ongoing at a site, the utility capacity available for the value chain analyzed may be restricted at some sites. For example, a wastewater treatment plant may only be capable of treating limited additional quantities of wastewater with the restriction being on a performance parameter such as total organic carbon content of the additional wastewater or the absolute quantities of wastewater generated. These kinds of restrictions can be integrated into the production network design model.

3.3.4 Plant Loading and Economies of Scale and Scope

So far, only scale effects related to capital expenditures have been discussed. Additionally, the allocation of products and production volumes to plants, often also referred to as *plant loading* decision, can be associated with scale effects in operating expenditures (cf. Cohen and Lee 1985, pp. 160-161). Generally, increasing production volumes at a plant leads to a spread of fixed costs such as plant infrastructure/management costs across larger volumes and hence reduces unit production costs. Variable production costs can show scale effects, too. For example volume discounts might be negotiated in materials or energy purchasing. Another type of scale effect in variable costs is referred to as *learning curves* (cf. Heizer and Render 2005, pp. 574-580). The idea behind the concept of learning curves is that processes are improved each time they are performed. The result is a cost curve that decreases with cumulative production volumes.

Also, product mix effects can affect production costs. Both synergies from producing certain products together at the same plant and additional complexity costs from producing an increasing number of products at the same plant, commonly referred to as *(dis-) economies of scope*, can be found. Diseconomies are for example caused by more frequent and more complex changeover processes which lead to capacity losses (cf. Mazzola 1989, pp. 923-924). A complex product mix can also increase costs in production planning, working capital requirements or material handling (cf. Billington and Davis 1992, p. 590; Cohen and Moon 1990, pp. 270-271; Hayes and Wheelwright 1984, pp. 90-117; Skinner 1974). Additionally, non-quantitative aspects such as supply risks play an important role in product-plant allocation decisions. For example, it is typically preferred to produce major products at two or more sites simultaneously to be able to react to supply disruptions. Examples such as the fact that during the 2003 political crisis in Venezuela many factories had to be shut down for up to 6 months and that in 2005 hurricane Katrina caused severe supply network disruptions in many industries in the U.S. illustrate the importance of this aspect.

Fixed cost effects are included in most production network design models but scale and scope effects related to variable costs and learning curve effects lead to concave cost functions (cf. Cohen and Moon 1990, p. 274). While these can be converted into piecewise linear cost functions, model complexity increases significantly both from a data preparation perspective (see Anderson (1995) for an approach to measure the impact on manufacturing overhead costs) and the mathematical solution process. Hence, most production network design models assume linear cost functions ignoring scale and scope effects related to variable costs.

Scale economies are for example considered by Cohen et al. (1989). The authors avoid mathematical non-linearities by modeling purchasing scale effects via discrete vendor contract options and production scale effects via fixed-cost charges and pseudo products (cf. Cohen et al. 1989, pp. 76-77). Syam (2000) uses piecewise linear cost functions both for labor and manufacturing costs. Verter (2002) actually employs non-linear optimization techniques. As a consequence, the simplifications employed, namely considering a single product, single period, single echelon domestic setting, render his model inappropriate for application to real-world production systems. Both Syam (2000) and Harkness and ReVelle (2003) further extend the idea of scale economies by modeling capacity "overutilization" via cost functions that, after exceeding a certain production volume, revert to assuming increasing unit production costs.

Since scope economies are especially hard to quantify, a separate class of optimization models solely dealing with plant loading decisions can be found. For example, Mazzola and Schantz (1997) propose a non-linear mixed integer program that combines a fixed cost charge for each plant-product allocation, a fixed capacity consumption to reflect plant setup and a non-linear capacity-consumption function of the total product portfolio allocated to the plant. To develop the capacity consumption function the authors build product families with similar processing requirements and consider effects from intra- and inter-product family interactions. Based on a linear relaxation the authors explore both tabu-search heuristics and branch-and-bound algorithms to obtain solutions.

In a model developed to analyze the trade-off between scale advantages from product-focused factories and reduced transport costs from market-focused factories Cohen and Moon (1991) and Moon (1989) consider a fixed charge incurred for each product-plant allocation and a concave production cost function. The cost function is transformed into a piecewise linear function. In a model developed for the paper industry, Philpott and Everett (2001, pp. 229-230) use pre-determined product mix "clusters" that are selected using binary variables and for which the effects on unit production costs and technical capacity are specified exogenously to model scope effects.

A few network design models explicitly cover scope effects, too. For example, Klincewicz and Luss (1987) include a fixed-cost charge for every product allocated to a facility. In a model developed for pharmaceutical industry, Papageorgiou et al. (2001) model capacity requirements for changeover processes. They allocate products to individual production lines and implicitly assume that only one campaign is produced per product and time period. Cohen and Moon (1990) combine fixed cost effects with a variable unit cost function that contains both additional scale and scope effects to analyze their effects and the effects of varying transportation costs on production network design using a single-period, non-linear model. They find that economies of scale and scope can have considerable effects on optimal production network design.

Benjaafar and Gupta (1999), while taking a machine perspective and not considering effects of geographically dispersed facilities, show that it is generally desirable to minimize the number of different products assigned to an individual facility. The heuristic they developed maintains a balanced utilization of facilities while minimizing the product sharing between facilities. According to Papageorgiou et al. (2001, pp. 275-277) this is especially relevant in pharmaceutical industry where complex product portfolios often lead to significant capacity losses because changeovers may result in equipment idle times of up to one month.

While fixed cost degression is an important aspect, in specialty chemicals production networks variable cost scale effects and learning curves were deemed to be of limited relevance. By centralizing purchasing of strategic raw materials a globally active company is often capable of negotiating volume discounts in materials purchasing even if production volumes are spread across multiple plants. Due to the nature of chemical production processes, efficiency is largely determined in the process development stage and improvements are generated via implementation of discrete process improvements instead of learning effects that can be mathematically linked to cumulative production volumes (cf. Sinclair et al. 2000). These process improvements can often be introduced at several plants within a relatively short period of time. To account for these peculiarities of chemicals production, the proposed model considers scale effects from fixed cost degression and discrete process improvements. The latter are incorporated via recipes and capacity consumption factors that are time-dependent, enabling the decision makers to include effects of expected process improvements.

However, diseconomies of scope typically occur in chemical industry due to changeover processes. Whether these effects have to be included at the network design level or whether including them at a later stage in a hierarchical planning approach, as for example described by Hax and Meal (1975), is sufficient, depends on the magnitude of the scale/scope economies existent with the type of products considered. As product mix effects were not a high-priority issue in the application areas considered in the course of this work and due to the problems inherent in estimating the underlying parameters, they are ignored in the basic model. Instead, the decision makers can define upper and lower bounds on the number of sites that are allowed to simultaneously produce a product. The idea behind this approach is to contain the complexity of the production network while at the same time eliminating the risks caused by producing important products only at a single site. This concept could be extended by requiring important products to be produced in several geographical areas or only in countries with less pronounced political risks. Since an explicit modeling of the cost effects associated with plant loading might be required for other value chains, an extension to include these effects is proposed in Chapter 3.4.3.

3.3.5 Specific Factors of Global Production Networks

As Verter and Dincer (1995, p. 265) point out, availability and cost of production factors such as labor, raw materials and manufacturing technologies vary strongly between countries. On top of that, global network design models have to include a range of factors related to international trade such as exchange rates or tariffs. Below, major aspects of international production network design are discussed.

Exchange Rates

The theory of (absolute) Purchasing Power Parity (PPP, cf. Eiteman et al. (2006, pp. 102-110) for details) states that under efficient market conditions exchange rates should be at an equilibrium where the price for all goods is identical in all countries (law of one price). Relaxing some of the assumptions, relative PPP states that in the long run exchange rate movements should be driven only by inflation rate differential. However, in reality non-traded goods, currency trades for capital asset transactions, central bank intervention, etc. affect exchange rates leading to significant deviations from the PPP exchange rates. The resulting currency fluctuations affect companies in different ways. Generally, one distinguishes between three types of currency exposure (cf. Eiteman et al. 2006, pp. 253-254; Dornier et al. 1998, pp. 307-308; Srinivasulu 1981, pp. 1-15):

- *Translation* or *accounting exposure* refers to the effects currency fluctuations can have on consolidated financial statements
- *Transaction exposure* refers to the effects on contractually defined cash flows and
- *Operation exposure* (also called economic exposure, competitive exposure or strategic exposure) refers to the change of the present value of a firm resulting from changes to future operating cash flows caused by unexpected currency fluctuations.

While the first two types of currency exposure can be controlled using financial market tools such as currency hedging, operating exposure requires a strategic response since financial instruments typically fail to reduce currency risks in the long run (cf. Copeland and Joshi 1996). Devising this strategic response is complicated by the fact that the effects of operating exposure are difficult to assess. Amongst other aspects the extent to which the company can adjust prices in reaction to exchange rate changes, the impact of the currency revaluation on raw material prices, demand and the competitive position of other market players have to be considered (cf. Eiteman et al. 2006, pp. 301-334; Dornier et al. 1998, pp. 311-314 or Lessard and Lightstone 1986). The consequences of exchange rate changes on the profitability of the business a production network serves can be profound. Mohamed (1999) shows, based on a simple example, that real exchange rate movements (change of nominal exchange rate adjusted for inflation levels in the countries involved, cf. Dornier et al.

1998, p. 309) can lead to a decrease of profits by almost 50% if capacity constraints do not allow the company to re-allocate production volumes within the network. Similarly, Vidal and Goetschalckx (1996, pp. 17-18) demonstrate that optimal network design solutions are strongly affected by exchange rates.

The key strategy to hedge against operating exposure is a global diversification of the company's operating and financing base (cf. Eiteman et al. 2006, pp. 310-321). Supply network design can contribute to this by matching currency cash flows for example via sourcing raw materials or components in foreign currencies, using risk-sharing agreements in international materials sourcing contracts or setting up operations in a foreign country (cf. Dornier et al. 1998, pp. 316-317). Maintaining excess capacity enables a company to shift production in reaction to exchange rate fluctuations. Thus, the company can actively exploit exchange rate changes (cf. Cohen and Huchzermeier 1999, p. 679; Dasu and Li 1997, pp. 716-717; Rosenfield 1996, pp. 332-333; Kogut 1985b, p. 33). In some cases managing operating exposure has been the sole or primary driver for changes to a production network. For example, both BMW and Mercedes Benz established production facilities in the U.S. to hedge against currency risks and Volkswagen's production activities in the U.S. were both established and closed primarily due to currency issues (cf. Kim and McElreath 2001, p. 23; Srinivasulu 1981, pp. 16-19).

Different approaches of how to integrate exchange rate uncertainty into manufacturing strategy have been proposed in literature. While a comprehensive review is not within the focus of this work, a few major contributions are presented below. Generally, it is possible to distinguish between approaches based on network flow models and those based on options valuation (cf. Huchzermeier and Cohen 1996, p. 101). The option value approach seeks to identify the additional value of operational flexibility from multinational supply networks that allow flexible shifting of production volumes or suppliers in response to exchange rate fluctuations. For example, Huchzermeier and Cohen (1996) assess the options value from both product design flexibility and manufacturing flexibility. They propose a hierarchical approach that feeds the results of a complex stochastic exchange rate model, alternative global manufacturing strategy options and transition costs into a supply network design evaluation model to determine the value of each product/network design option for each exchange rate scenario. The options value of the available degree of flexibility is then obtained by calculating the difference between the case where no switching between options is allowed and the case where switching is permitted. However, according to Lun and Peske (2002) the real options valuation approach is rarely used in practice due to its complexity. Integrating it into a comprehensive network design model further aggravates this problem. Therefore, it will not be considered here but the reader is instead referred to examples in literature (e.g., Cohen and Huchzermeier 1999; Huchzermeier and Cohen 1996 or Kogut and Kulatilaka 1994). A general exposition of the real options approach can be found in Dixit and Pindyck (1995) or corporate finance textbooks.

The effects of exchange rate fluctuations on optimal production network design can also be assessed without explicitly calculating the economic value of flexibility options. Lowe et al. (2002) suggest a two-phase screening approach based on discrete exchange rate scenarios. The first phase assesses the opportunity costs of keeping the status quo for a specified period (e.g., one year) using a limited data set. If the opportunity costs indicate that action is required, alternative network configurations are generated. These alternative configurations are assessed by calculating minimum costs for every exchange rate scenario. A subsequent filtering process based on different criteria such as pareto-optimality, maximum regret, mean-variance, stochastic dominance or pairwise stochastic comparison is employed to identify "good" configurations. For configuration options that passed the screening in phase 1 a comprehensive multi-period analysis is performed in the second phase to identify the preferred option.

Gutierrez and Kouvelis (1995) develop what they call a robustness approach to international sourcing. Their objective is to identify an international supply network that possesses reasonable costs under any exchange rate scenario considered likely for the planning horizon. To this end they compare for each design option the performance under all exchange rate scenarios considered relative to the optimal solution for this scenario. A design option is deemed to be robust if a pre-specified upper bound for the maximum percentage regret is not exceeded. The logic of their approach can also be transferred to production network design. It should be noted that in this approach it is assumed that the network design chosen cannot be changed in later periods of the planning horizon.

Since the analysis for specialty chemicals production networks has to be performed at the value chain level the options to holistically assess the effects of exchange rate risks are limited. The effects currency fluctuations have on the company can only be assessed properly if all cash flows are taken into account. Nevertheless, management typically expects each value chain's business to be viable by itself. Integrating currencies into the model can provide valuable information on the distribution of cash flows across currencies and the resulting degree of operating exposure within the value chain. To this end all cost items are modeled in the currency they are actually denominated in. The strategic planning process can be supported using simple "what if..." analyses to understand the potential effects of exchange rate fluctuations (cf. Cohen and Lee 1989, pp. 96-98). Including a restriction on the spread of cash flows across currencies that is gradually tightened to balance cash flows can provide a cost assessment for (partially) eliminating operating exposure (cf. Meyer 2005, p. 166). Additionally, design alternatives that are more robust towards exchange rate changes (cf. 3.4.4), e.g., via providing overcapacity, can be identified. The effects of exchange rate fluctuations will be further explored in the application case study in Chapter 5.5.2.

Duties and Duty Drawback[32]

Duties are import taxes levied at the border for importing a product into the respective country. Duties are usually defined as a percentage of the value of imported goods including freight costs to border of destination country/trade area (see Appendix 2 for further details on duty procedures) with product-specific rates commonly varying between 0 and 10% and in extreme cases exceeding 30%. The applicable tariff rate for importing a product depends not only on the product but also on the country of origin. Within free trade zones no duties have to be paid and for many trade relations preferential rates apply either due to bilateral trade agreements or because of exemptions to support less developed countries. Several options exist in production network design to reduce tariff costs. In some cases, duties can be avoided by choice of production locations. In other cases, refunds on duties paid (duty drawbacks) can be obtained. Figure 19 gives an overview of duty avoidance options and duty drawbacks.

Models from literature differ with respect to the extent to which they cover the complexity of tariff structures. Simplified modeling approaches only consider duties levied for supplying a certain market via imports or simply integrate tariffs into transportation costs (e.g., Kouvelis et al. 2004, p. 129). As a basis for calculating the tariff sales prices (e.g., Haug 1985, p. 86), transfer prices (e.g., Canel and Khumawala 1996, p. 56; Vidal and Goetschalckx 2001, p. 145), transfer prices plus transportation costs (Canel and Khumawala 1997, p. 1898) or unit manufacturing costs (e.g., Chakravarty 1999, p. 4241; Bhutta et al. 2003, p. 204) are used. Duty drawbacks are only covered by a few models (e.g., Arntzen et al. 1995, pp. 88-90; Meyer 2005, pp. 136-138). Meyer uses a simplified drawback model based on pre-defined duty and duty drawback values for each product. He allows netting out across different products (drawbacks can be obtained also for exported products that do not contain any of the previously imported parts), which is usually not allowed by tax regulations. Arntzen et al. com-

[32] This chapter is based on McDonald (1997), pp. 65-79 and Jackson (1997).

prehensively model duties and duty drawbacks based on a global bill of materials but acknowledge that tracing back the lineage for purposes of duty drawback might in practice only be possible for a limited segment of the product tree. Chakravarty (2005) determines optimal tariff-adjusted pricing strategies via overhead cost allocation.

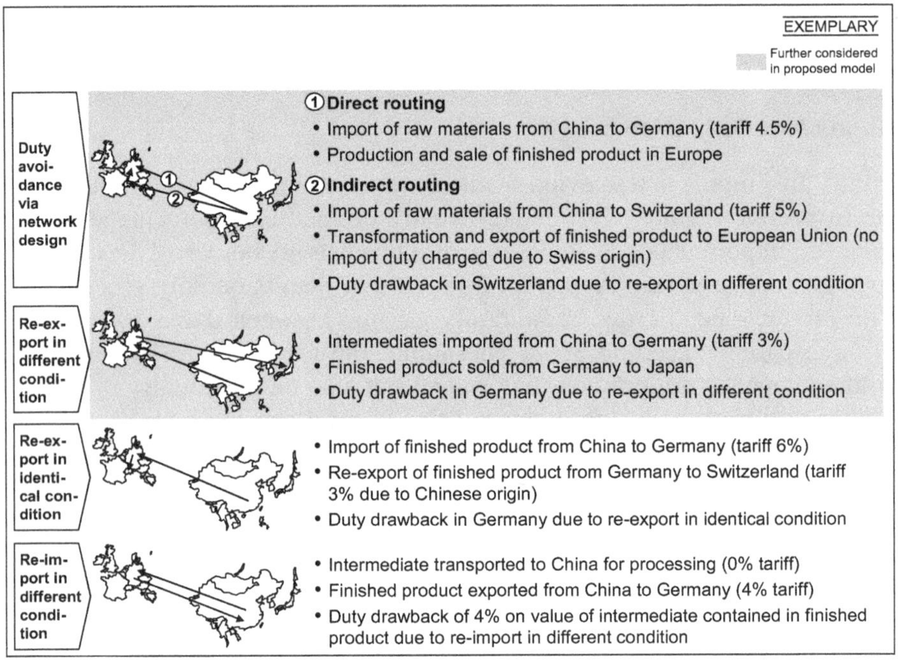

Fig. 19. Duty avoidance and duty drawback options

For specialty chemicals, tariffs often constitute a higher share of total costs than transportation costs making obvious the need to explicitly model tariff regimes. However, from the four duty-related issues shown in Figure 19 only the first two have to be included in the model. Re-exports in identical condition typically occur in distribution networks. Re-imports in different condition are often linked to outsourcing of certain production steps to a toll manufacturer. For example, a product may be exported to a low-cost country for certain personnel-intensive production steps and the processed product is later re-imported. In specialty chemicals this is an exception. To model duty drawbacks, the lineage could theoretically be traced for multiple production levels along the bill of materials. To reduce complexity, the proposed model covers only raw material duty drawbacks for re-export after one production level or after two production levels taking place at the same site. Drawback for intermediates imported from another site is limited to re-exports after one production level.

Non-tariff Trade Barriers

According to Lee and Swagel (1997, p. 372) and Anderson and Schmitt (2000, p. 3) with tariffs being reduced in the context of WTO/GATT negotiations and free trade agreements, non tariff trade barriers are increasingly used to protect domestic markets. The most common ones are local content rules, import quotas, subsidies or technical barriers. Especially local content rules, due to their importance in industries such as the automotive industry (cf. Munson and Rosenblatt 1997, pp. 278-279) have been included in supply chain models. For example, the model proposed by Arntzen et al. (1995, pp. 88-89) contains them and Munson and Rosenblatt (1997) have developed models to incorporate local content rules into sourcing decisions both in single plant and multiple plant production networks. However, these kinds of trade restrictions are of minor importance in specialty chemicals and will not be further considered.

International Taxation and Transfer Prices

As, amongst others, Choi et al. (2002, p. 472) and Hines Jr. (1999) point out, international tax regimes have a strong influence on where to invest, how to organize a business and where to recognize elements of revenues and costs. In order to reduce the overall tax burden, companies seek to allocate expenses to high-tax countries and profits to low-tax countries. According to a survey conducted by Ernst & Young (2003), multinational companies consider transfer prices to be the most important issue in international taxation. Nieckels (1976, pp. 142-146) demonstrates that transfer price changes can significantly affect the after-tax profit of a company. Similarly, the allocation of internal transport costs to the sending or receiving country can be used to allocate profits to favorable locations (cf. Cohen and Lee 1989, p. 90). As a consequence, supply network design, taxes and transfer pricing are ideally considered in an integrated approach (cf. Vidal and Goetschalckx 2001, p. 135; Murphy 1998).

In order to incorporate tax regimes into supply network design, the respective model has to maximize after-tax cash flows and be able to allocate transportation costs and determine optimal transfer prices. Determining optimal transfer prices is the most complex aspect because tax authorities have adopted strict rules regulating transfer pricing in order to block the exploitation of tax differentials and tax authorities from at least two countries are involved (cf. Hollas 2004, p. 49).[33]

While several global supply network design models explicitly include the effects of taxes based on externally determined transfer prices (e.g., Papageorgiou et al. 2001), only a few models simultaneously determine/optimize transfer prices. Nieckels (1976) proposed a non-linear model, which does not include location decisions, that is solved iteratively, combining a heuristic solution procedure to assign values to the transfer prices and the optimization of a linear flow optimization program based on the transfer prices obtained with the heuristic. The model proposed by Cohen et al. (1989) is also non-linear. A decomposition procedure is used to obtain a solvable linear model. While this procedure does not lead to a global optimum, it is reiterated until no further significant improvements are achieved. Canel and Khumawala (1997 and 1996) include transfer price decisions in a single-product optimization model. They fix the transfer price to either the lower or upper bound determined by tax regulations before solving the remaining optimization problem. Vidal and Goetschalckx (2001) develop a comprehensive non-linear model to optimize transfer prices in global logistics systems. However, the model does not include network design elements. A heuristic solution procedure and an approach to solve the problem within a pre-defined optimality gap are proposed.

Martel et al. (2005, pp. 26-27) exploit the fact that transfer prices can often only be chosen within tight regulatory constraints to preserve the linearity of their model while still optimizing strategic network design and transfer prices simultaneously. Instead of treating transfer prices as continuous variables, discrete changes of the markup percentage are allowed for a country's transfer prices. A binary decision variable is included to decide which multiplier is used. According to the authors the approach generates good results if the number of markup alternatives is limited.

The model proposed in this work seeks to optimize the production network of individual value chains and hence neither covers all production activities taking place in a country nor at a single site simultaneously. Due to the lack of taxation-relevant data implied by this value chain perspective, pre-tax cash flows are used in the basic model but an extension to include the effects of taxes is presented in Chapter 3.4.3. Fleischmann et al. (2006) took the same decision in a comparable situation.

---

[33] For further details on transfer pricing refer to Feinschreiber and Kent (2003), Choi et al. (2002, pp. 491-505) or Abdallah (1989). Feinschreiber (2004) provides details on the different transfer pricing methods and Feinschreiber (2000) discusses transfer pricing regulations of more than 40 countries.

Similarly, the degree of freedom available to optimize transfer prices depends on factors outside the model such as accumulated deficits or the profitability of other business activities in a country. Furthermore, legal constructs such as principal trading companies (cf. Murphy 1998) are employed in practice to optimally allocate profits within a company. Finally, from a modeling perspective the simultaneous optimization of network design and transfer prices leads to a non-linear model (cf. Schmidt and Wilhelm 2000, p. 1510; Verter and Dincer 1995, p. 278) that is difficult to solve to optimality for realistic problem instances. Therefore, transfer prices are not optimized but instead determined outside the optimization model. Furthermore, transfer prices are assumed to be independent of the transfer destination because tax authorities in the country of origin usually do not accept differentiated transfer prices. In literature differentiated approaches can nevertheless be found as well (e.g., Kouvelis et al. 2004, p. 130).

Government Incentives

In some cases host governments provide investment incentives to attract certain industries or to develop certain areas. These can take different forms such as tax credits, investment subsidies, export incentives, subsidized credits or training grants. Many international supply network design models consider such incentives. For example, Canel and Das (2002, p. 113) include direct investment subsidies while Hodder and Dincer (1986, p. 603) and Kouvelis and Rosenblatt (2002, p. 253) include subsidized loans. Haug (1985, p. 90) considers training grants provided for hiring new employees and Canel and Khumawala (1996, p. 58) include export incentives in the form of a cash payment. In specialty chemicals investment subsidies generally do not play an important role. In order to be able to accommodate government incentives, an investment subsidy factor based on the total capital investment expenditures is included in the model. If required, other types of incentives such as training grants can be incorporated, too.

Political Risk

While being of qualitative nature, political risks are an important aspect of production network design. In a relatively simple approach, Syam (2000, pp. 175-176) incorporates a risk reduction strategy based on regional diversification. He limits the number of sites that can be operated in each geographical region, effectively forcing the model to spread production across regions. Syam assumes these limits to be predetermined and does not discuss how to derive reasonable limits. Alternatively, he suggests to use risk indices (these are again not further specified) and illustrates the cost savings vs. increased risk trade-offs that can be made by evaluating alternative limits on cumulated risk.

Global specialty chemicals companies already operate plants in many different countries and have the expertise to navigate in the respective business environments. Therefore, typically political risks are not incorporated into quantitative models but considered when selecting the potential investment candidate countries. If desired, the model can be extended to include political risk based on an aggregate risk parameter to analyze risk/return profiles for alternative network configurations. For a more detailed discussion of the major elements of political risk see Appendix 3.

3.3.6 Single Sourcing

Single-sourcing restrictions can be employed between different echelons of a supply chain. Their objective often is to reduce the complexity of the supply network. For example, in a production-distribution model a restriction might be included to ensure that each customer is only served by one distribution center (e.g., Tsiakis et al. 2001, p. 3590; Geoffrion and Graves 1974, p. 823).

In specialty chemicals industry another motivation for single-sourcing restrictions exists. While product batches originating from different sites are identical insofar as that they comply with the relevant specifications, parameter/quality differences exist between plants. Supplying regular customers from more than one plant can lead to increased variability of product characteristics and have negative effects on the customer's production processes. For this reason it may be desirable to enforce single-sourcing restrictions in some markets. As only a subset of the customers are usually affected by this problem, it is normally sufficient if a certain percentage of a market's demand is met from a single site throughout at least a part of the planning horizon. This can be achieved either by directly specifying a certain quantity to be supplied from a certain plant to the respective market (e.g., in the case of corresponding contractual obligations or existing allocations). Alternatively, a formulation that allows the model to select the plant best suited to supply the market can be used. Following the same process stability logic, it is also desirable that a plant using internally produced intermediates is not supplied from several plants in parallel but instead dedicated internal supplier relationships are established. Finally, certain plant-market allocations may be excluded (negative list) to reflect political sensitivities or trade restrictions (e.g., India and Pakistan).

3.3.7 Product Transfers

Even if the receiving plant has been running the equipment required to produce a certain product for a long period of time chemical products cannot be transferred easily from one site to another. Instead, recipes have to be adjusted, personnel has to be trained and trial batches have to be produced. However, only a few models in literature consider these aspects. Lee (1991, p. 169) includes fixed setup costs to create the capability to handle a product at a facility. For pharmaceutical industry Papageorgiou et al. (2001, p. 277) include product scale up further distinguishing between the actual scale-up and qualification runs required to demonstrate the capability of meeting specifications to regulatory agencies. The proposed model considers both external costs associated with scale-up and the costs/capacity consumption induced by the trial runs.

3.3.8 Other Model Features

Some models also include transportation mode selection (e.g., air freight, ship or truck transport, full container load or less than container load) either solely based on cost comparisons or also considering the trade-off between transportation costs and pipeline inventory or lead time (cf. Zeng 2002; Vidal and Goetschalckx 2000, pp. 106-107; Jayaraman 1998, pp. 474-476). Due to the limited importance of transport costs in specialty chemicals and the fact that products are typically transported between plants and markets in full container loads this level of detail does not provide additional insights.

While raw material supply is usually taken for granted, some models also include vendor selection, vendor capacity constraints or vendor delivery reliability (e.g., Martel et al. 2005; Vidal and Goetschalckx 2000; Vidal and Goetschalckx 1996).

3.4 Mathematical Optimization Model

In this chapter the MILP model that incorporates the aspects discussed in the previous chapter is presented. Chapter 3.4.1 introduces the notation used and Chapter 3.4.2 presents and discusses the model formulation. Extensions to the basic model are presented in Chapter 3.4.3.

3.4.1 Model Notation

Due to the various economic aspects that have to be captured by the optimization model, a large number of indices, parameters and decision variables is required. The notation used is contained in Table 5 (indices), Table 6 (parameters), Table 7 (calculated parameters) and Table 8 (decision variables). Additionally, the link of major parameters and decision variables to the production network and the cost structure underlying the model are visualized in Figures 20 to 23. To simplify the notation some parameters are defined at a higher aggregation level (e.g., exchange rates per currency or transport costs at source-destination country level) and the applicable link is indicated in brackets after the index (e.g., $exrate_{q(rs)t}$ indicates the exchange rate for the currency raw material r is denominated in at site s).

Table 5. Indices used in model formulation

Primary index sets			
$c \in C$	Countries	$q \in Q$	Currencies
$f \in F$	Plants	$r \in R$	Raw materials
$g \in G$	Plant classes	$s \in S$	Sites
$p \in P$	Products	$t \in T$	Time periods
Induced index sets			
CE_s	Countries that are export destinations relative to site s		
CNE_s	Countries that are no export destination relative to site s		
F_g	Plants belonging to plant group g		
F_p	Plants that are capable of producing product p		
FSA	Set of existing and potential plant-site combinations		
FSE	Plant-site combinations existing at beginning of planning horizon		
FSN	Potential plant-site combinations		
PE	Final products (only external demand possible)		
PI	Intermediate products (internal and external demand possible)		
P_f	Products that can be produced at plant f		
P_i	Products requiring intermediate product i		
$PFSE$	Product-plant-site allocations existing at beginning of planning horizon		
PSC^-	Single-sourcing negative list (infeasible product-site-country allocations)		
PSC^+	Single-sourcing positive list (required product-site-country allocations)		
R_p	Raw materials required for production of product p		
SI_s	Sites that are an import source relative to site s		
SE_s	Sites that are an export destination relative to site s		
SNE_s	Sites that are no export destination relative to site s		
SU	Site-utility combinations with capacity restrictions		

Table 6. Production system and financial parameters

Production system parameters	
$capreq_{pst}$	Capacity consumption factor for product p at site s in period t
$contvol_p$	Quantity of product p fitting into one standard container (40 ft)
$curFTE_{fs}$	FTEs employed at plant f at site s at beginning of planning horizon
$existmod_{fs}$	Additional production lines at plant f at site s at beginning of horizon
$fixFTE_{fs}$	Fixed FTEs required to operate plant f at site s
$intreq_{pist}$	Quantity of intermediate i req. per unit of product p at site s in period t
$minsupply_{psct}$	% of demand for product p in country c to be supplied from site s in t
$mtcap_{fs}$	Capacity per production line of plant f at site s
$\overline{mtcap}_{fs}$	Maximum technical capacity of plant f at site s
$\underline{mtcap}_{fs}$	Minimum technical capacity of plant f at site s
$rreq_{prst}$	Quantity of raw material r req. per unit of product p at site s in period t
$scupstat_{pfs}$	Scale-up status of product p at plant f at site s at beginning of horizon
$scupvol_p$	Production volume required for trial runs to scale up product p
$\overline{S}_p$	Maximum number of sites producing product p per time period
$\underline{S}_p$	Minimum number of sites producing product p per time period
$utcap_{sut}$	Capacity of utility (utility parameter) u at site s in period t
$utreq_{pust}$	Quantity of utility u required per unit of product p at site s in period t
$varFTE_{fs}$	Number of FTEs required per production line of plant f at site s
Financial and market parameters	
$attr_s$	Personnel fluctuation rate at site s in period t
$capexreq_{fst}$	Expenditure for setting up plant f at site s in period t
$closcost_{fst}$	Closure costs for plant f at site s in period t
$demand_{pct}$	Demand for product p in country c in period t
$exrate_{qt}$	Exchange rate of currency q against home currency in period t
$fixcost_{fst}$	Fixed costs for operating plant f at site s in period t
$FTEcost_{st}$	Personnel costs at site s in period t
inc_{st}	Investment subsidy in % of capital expenditures at site s in period t
$invcc_t$	Inventory carrying costs in % of product value in period t
$linecapex_{fst}$	Expenditure per additional production line of plant f at site s in period t
$linefixcost_{fst}$	Fixed costs per additional line of plant f at site s in period t
$othervarcost_{pst}$	Other variable costs of product p at site s in period t
$price_{pct}$	Price of product p in country c in period t
PVF_t	Present value factor for period t
rpr_{rst}	Landed costs (excl. tariff) of raw material r at site s in period t
$salescost_{ct}$	Sales costs in country c in period t in % of revenues

Table 6 (continued). Production system and financial parameters

$scupcost_{p}$	Cost of setting up production of product p (in home currency)
$sevfactor_{s}$	Severance payment factor at site s as multiple of annual FTE costs
$tariff_{pc'c}$	Tariff rate for product p if imported from country c' into country c
$trcost_{c'ct}$	Freight rate for a 40ft container from country c' to country c in period t
$trdur_{cc'}$	Transport duration from country c to country c' in days
$trprice_{pst}$	Transfer price of product p originating from site s in period t

Table 7. Calculated parameters and modeling parameter

Calculated parameters	
$capex_{st}$	Capital expenditures incurred at site s in period t
$capex_{fst}$	Capital expenditures incurred at plant f at site s in period t
$cfix_{st}$	Fixed costs incurred at site s in period t (only for new sites)
$closurecost_{st}$	Closure costs incurred at site s in period t
$closurecost_{fst}$	Closure costs incurred at plant f at site s in period t
$cvar_{st}$	Variable costs incurred at site s in period t
$distrcost_{ct}$	Distribution costs incurred in country c in period t
$facFTE_{fst}$	Number of FTEs required at plant f at site s in period t
$imptariff_{ct}$	Import tariff costs incurred in country c in period t
$inventcc_{t}$	Inventory carrying costs incurred in period t
$itariff_{is'st}$	Duty incurred at site s for import of intermediate i from site s' in period t
$netrev_{ct}$	Net revenues realized in country c in period t
$rtariff_{st}$	Duties incurred at site s for import of raw materials in period t
$scupcosts_{st}$	Product scale-up costs incurred at site s in period t
$scupdem_{pfst}$	Trial quantities for scale-up of product p at plant f at site s in period t
$secdemand_{pst}$	Secondary demand for intermediate p at site s in period t
$sevpay_{st}$	Severance payments incurred at site s in period t
$shipcost_{st}$	Internal transport costs incurred at site s in period t
$tcap_{fst}$	Technical capacity of plant f at site s in period t
$ttariff_{st}$	Total tariff payments incurred at site s in period t
Modeling parameter	
$BigM$	Sufficiently large positive number

Table 8. Decision variables

Decision variables	
$dext_{psct}$	Distribution volume of product p from site s to country c in period t
$dint_{pss't}$	Distribution volume of intermediate p from site s to s' in period t
$prvol_{pfst}$	Production volume of product p at plant f at site s in period t
u_{pfst}	1, if product p has been scaled up at plant f at site s in period t; 0 else
v_{pst}	1, if product p is produced at site s in period t; 0 else
w_{fst}	1, if existing plant f at site s has been expanded; 0 else
$x_{pss't}$	1, if intermediate p is supplied to site s' from site s in period t; 0 else
y_{fst}	1, if plant f is open at site s in period t; 0 else
z_{fst}	No. of additional production lines open at plant f at site s in period t

The major product flow variables and cost parameters associated with the material flows throughout the production network are also shown in Figure 20. Based on the assumption that a supplier with lowest landed costs is selected outside the model per raw material and site, raw material flows are not modeled explicitly. Supplier selection could however also be included in the model analogous to the extension to include make-or-buy decisions for intermediate products (cf. Chap. 3.4.3).

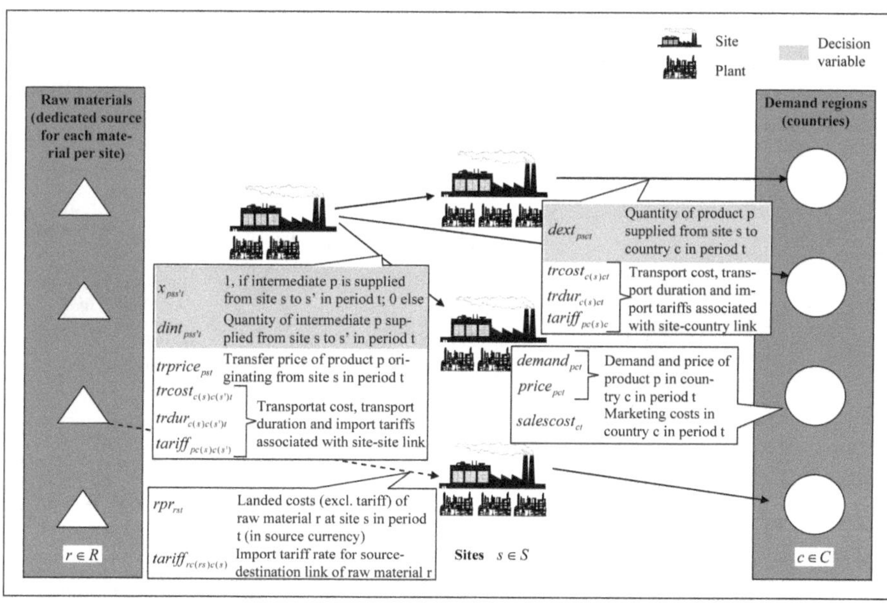

Fig. 20. Key parameters and decision variables at network level

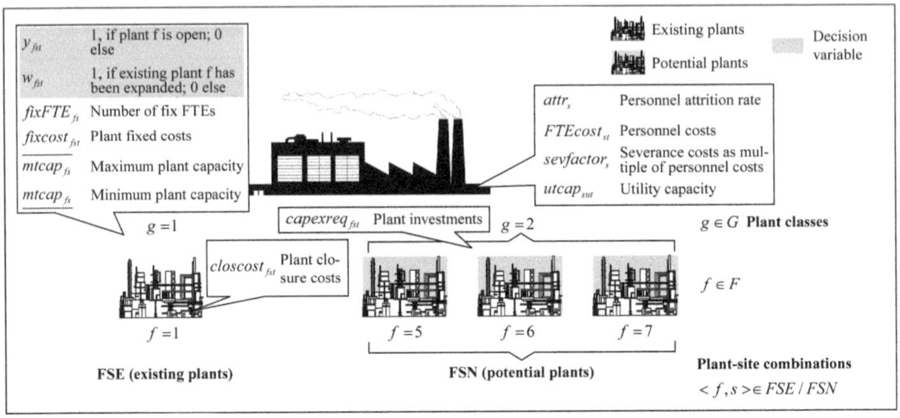

Fig. 21. Key parameters and decision variables at site level

Figure 21 visualizes central model elements at the site level. A site consists of infrastructure units (which are not modeled explicitly) and one or more plants belonging to different plant groups (of which some might not be modeled because they belong to different value chains). Additionally, a set of potential plants (belonging to a facility group not yet operated at the site) that could be constructed might be available and an expansion of existing facilities might be possible.

Within a plant (cf. Fig. 22) the major decisions concern product allocation, product transfers, production volumes and capacity changes.

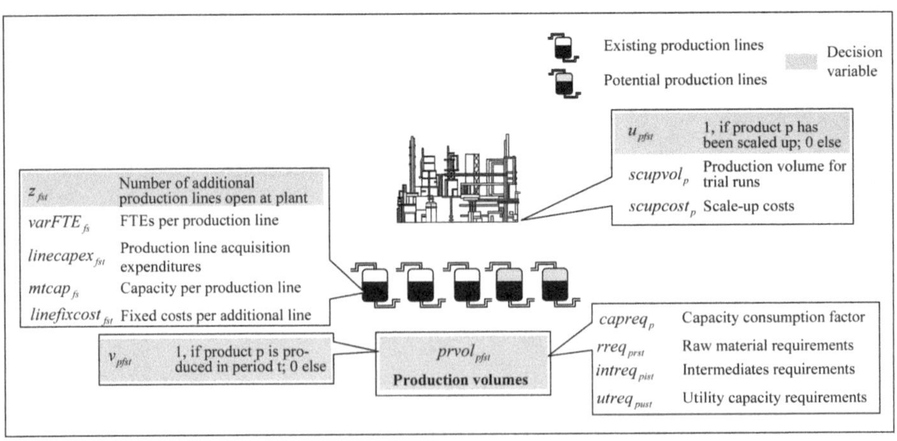

Fig. 22. Key parameters and decision variables at plant level

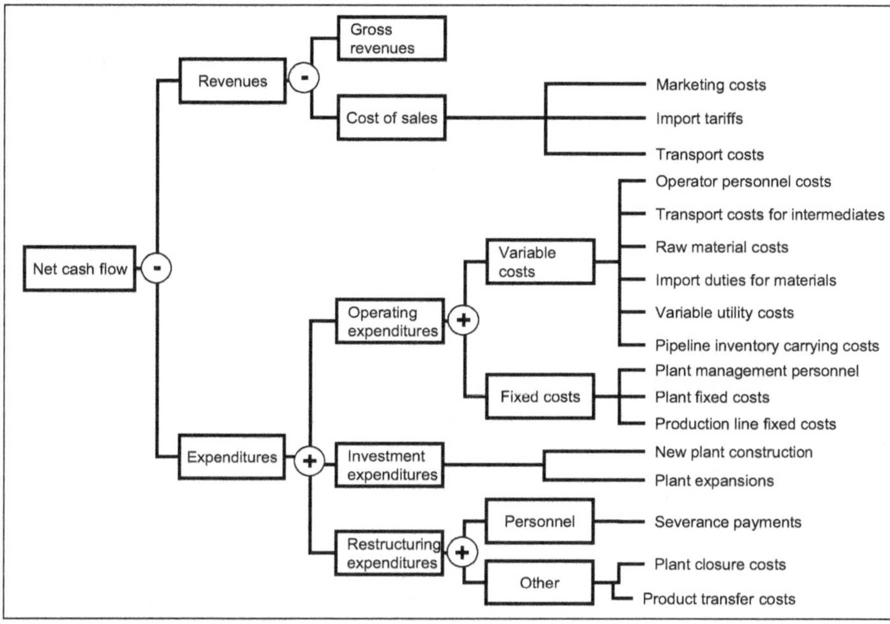

Fig. 23. Cost structure of optimization model

3.4.2 Model Formulation

Below, the basic model is presented and the formulation is discussed.

Objective Function

$$max \sum_{t \in T} \left[\begin{array}{l} \sum_{c \in C} \left[\begin{array}{l} netrev_{ct} \\ - imptariff_{ct} \\ - distrcost_{ct} \end{array} \right] \cdot exrate_{q(c)t} \\ - \sum_{s \in S} \left[\begin{array}{l} cvar_{st} \\ + shipcost_{st} \\ + cfix_{st} \\ + ttariff_{st} \\ + capex_{st} \\ + scupcosts_{st} \\ + closurecost_{st} \\ + sevpay_{st} \end{array} \right] \cdot exrate_{q(s)t} \\ - inventcc_t \end{array} \right] \cdot PVF_t \tag{3.2}$$

The objective function maximizes the net present value of cash flows before taxes. It contains three major components: country revenues, site costs and inventory carrying costs. To improve legibility, the equations calculating the parameters contained in the objective function are discussed below ahead of the actual model restrictions.

Cost Components of the Objective Function

$$netrev_{ct} = \sum_{p \in P} \sum_{s \in S} dext_{psct} \cdot price_{pct} \cdot (1 - salescost_{ct}) \qquad \forall c \in C, t \in T \qquad (3.3)$$

Net revenues (3.3) are calculated by adjusting gross revenues for sales costs incurred in a country. These sales costs include costs for marketing, country management, etc. and are expressed in per cent of the sales price.

$$distrcost_{ct} = \sum_{p \in P} \sum_{s \in S} dext_{psct} \cdot \frac{trcost_{c(s)ct}}{contvol_p} \cdot \frac{1}{exrate_{q(c)t}} \qquad \forall c \in C, t \in T \qquad (3.4)$$

Equation (3.4) calculates the transportation costs to deliver products from the production site to the market. Transport costs are assumed to be denominated in the home currency of the company and hence converted into the currency of the destination country for allocation purposes.

$$imptariff_{ct} = \sum_{p \in P} \sum_{s \in S} dext_{psct} \cdot \frac{trprice_{pst} + \frac{trcost_{c(s)ct}}{contvol_p}}{exrate_{q(c)t}} \cdot tariff_{pc(s)c} \qquad \forall c \in C, t \in T \qquad (3.5)$$

The import tariffs associated with the product flows between sites and markets are calculated in equation (3.5). The sum of transfer price and transportation costs is used to value products for tariff calculation. While in reality only the transportation costs to the border of the country are included in the product valuation for tariff purposes, this assumption was made to simplify the calculation.[34]

[34] As transport costs within the country of destination only make up approximately a third of total transport costs it was not deemed worthwhile to further split up the transport costs to reflect the fact that only freight costs until the product reaches the border of the destination country have to be included.

$$cvar_{st} = \sum_{p \in P} \sum_{f \in F} prvol_{pfst} \left[\begin{array}{l} \sum_{r \in R_p} rreq_{prst} \cdot rpr_{rst} \cdot \dfrac{exrate_{q(rs)t}}{exrate_{q(s)t}} \\ + \dfrac{capreq_p}{mtcap_{fs}} \cdot varFTE_{fs} \cdot FTEcost_{st} \\ + othervarcost_{pst} \end{array} \right] \qquad \forall s \in S, t \in T \qquad (3.6)$$

The variable costs incurred at a site are calculated in equation (3.6). Variable costs include costs for raw materials and utilities (modeled as raw materials), personnel costs and other variable costs. Each raw material is assumed to be supplied from a single supplier at each site. Since country of origin and currency denomination are often not identical, the invoicing currency is specified for each raw material.

$$shipcost_{st} = \sum_{p \in PI} \sum_{s' \in S \setminus \{s\}} \frac{dint_{ps'st}}{contvol_p} \cdot \frac{trcost_{c(s')c(s)t}}{exrate_{q(s)t}} \qquad \forall s \in S, t \in T \qquad (3.7)$$

Internal transport costs for intermediates are calculated analogous to distribution costs and are allocated to the receiving site. For intermediates produced at the destination site transport costs are assumed to be zero.

$$cfix_{st} = \sum_{f \in F} \left[\begin{array}{l} fixcost_{fst} \cdot y_{fst} \\ + linefixcost_{fst} \cdot z_{fst} \\ + fixFTE_{fs} \cdot FTEcost_{st} \cdot y_{fst} \end{array} \right] \qquad \forall s \in S, t \in T \qquad (3.8)$$

The fixed costs incurred at each site (3.8) include a fixed cost block for the plant itself, a fixed-cost charge for each additional production line open at the plant and the volume-independent personnel required to operate the plant.

$$ttariff_{st} = \sum_{i \in PI} \sum_{s' \in S \setminus \{s\}} itariff_{is'st} + rtariff_{st} \qquad \forall s \in S, t \in T \qquad (3.9)$$

Tariff costs for imported intermediates and raw materials are allocated to the receiving site. The detailed calculations of the tariffs are contained in equations (3.47) and (3.48).

$$capex_{fst} \geq \left[\begin{array}{l} (y_{fst+1} - y_{fst}) \cdot capexreq_{fst} \\ + (z_{fst+1} - z_{fst}) \cdot linecapex_{fst} \end{array} \right] \qquad \forall < f, s > \in FSA, t \in T \setminus \{n\} \qquad (3.10)$$

$$capex_{st} = \sum_{f \in F} capex_{fst} \cdot (1 - inc_{st}) \qquad \forall s \in S, t \in T \setminus \{n\} \quad (3.11)$$

The total investment expenditures incurred at a site have to be calculated in two steps. Equation (3.10) calculates the investments per plant. These are aggregated to the site level and adjusted for government investment incentives, defined as percentage of total investments, in equation (3.11). Investment expenditures are allocated to the time period preceding the commissioning of the technical capacity. A non-negativity constraint (3.54) ensures that plant/production line shutdowns do not lead to negative investment expenditures.

$$scupcosts_{st} \geq \sum_{p \in P} \sum_{f \in F} \left(u_{pfst} - u_{pfst-1}\right) \cdot \frac{scupcost_p}{exrate_{q(s)t}} \qquad \forall s \in S, t \in T \setminus \{1\} \quad (3.12)$$

$$scupcosts_{st} \geq \sum_{p \in P} \sum_{f \in F} \left(u_{pfst} - scupstat_{pfs}\right) \cdot \frac{scupcost_p}{exrate_{q(s)t}} \qquad \forall s \in S, t = 1 \quad (3.13)$$

Equations (3.12) and (3.13) calculate the costs caused by transferring production of a product to another plant (scale-up). These costs are incurred in the form of expatriates, trainings, etc. The costs of production trials are contained in the variable production costs of the site via equations (3.22/3.23) and (3.40/3.41).

$$closurecost_{fst} \geq \left(y_{fst-1} - y_{fst}\right) \cdot closcost_{fst} \qquad \forall < f, s > \in FSE, t \in T \setminus \{1\} \quad (3.14)$$

$$closurecost_{st} = \sum_{f \in F} closurecost_{fst} \qquad \forall s \in S, t \in T \setminus \{1\} \quad (3.15)$$

Closure costs for existing plants incurred at a site are again determined in two steps. Plant closure costs are calculated in equation (3.14) which, in combination with the non-negativity constraint (3.57), ensures that no negative closure costs are calculated. Closure costs are allocated to the time period during which the capacity is decommissioned. To account for lead times in closure decisions no plant closures can occur in the first period of the planning horizon. Equation (3.15) aggregates the closure costs to the site level.

$$sevpay_{st} \geq \sum_{f \in F} \left[\begin{array}{l} \left(facFTE_{fst-1} \cdot (1 - attr_s) - facFTE_{fst}\right) \\ \cdot FTEcost_{st} \cdot sevfactor_s \end{array} \right] \qquad \forall s \in S, t \in T \setminus \{1\} \quad (3.16)$$

$$sevpay_{st} \geq \sum_{f \in F} \left[\begin{array}{l} \left(curFTE_{fs} \cdot (1 - attr_s) - facFTE_{fst} \right) \\ \cdot FTEcost_{st} \cdot sevfactor_s \end{array} \right] \qquad \forall s \in S, t = 1 \quad (3.17)$$

The calculation of severance payments in equations (3.16) and (3.17) considers the natural fluctuation rate and personnel requirements as specified in equation (3.49). Again, non-negativity constraint (3.72) ensures that in case of personnel increases no negative severance payments are calculated. The same type of formulation could also be used to model training costs for newly hired employees but this was considered to be of minor relevance in the application cases pursued so far.

$$inventcc_t = \left[\begin{array}{l} \sum_{p \in P} \sum_{s \in S} \sum_{c \in C} dext_{psct} \cdot \frac{trdur_{c(s)c}}{360} \cdot trprice_{pst} \cdot invcc_t \\ + \sum_{p \in PI} \sum_{s \in S} \sum_{s' \in S} dint_{pss't} \cdot \frac{trdur_{c(s)c(s')}}{360} \cdot trprice_{pst} \cdot invcc_t \end{array} \right] \qquad \forall t \in T \quad (3.18)$$

Based on the average transport duration, inventory carrying costs for pipeline inventory are calculated (3.18). Further safety stocks are considered to be independent of the chosen network design and hence not considered.

Flow Constraints

$$\sum_{s \in S} dext_{psct} \leq demand_{pct} \qquad \forall p \in P, c \in C, t \in T \quad (3.19)$$

$$\sum_{s' \in S} dint_{ps'st} = secdemand_{pst} \qquad \forall p \in PI,\ s \in S, t \in T \quad (3.20)$$

$$secdemand_{ist} = \sum_{p \in P_i} \sum_{f \in F} prvol_{pfst} \cdot intreq_{pist} \qquad \forall i \in PI, s \in S, t \in T \quad (3.21)$$

$$\sum_{f \in F_p} prvol_{pfst} = \sum_{c \in C} dext_{psct} + \sum_{f \in F_p} scupdem_{pfst} \qquad \forall p \in PE, s \in S, t \in T \quad (3.22)$$

$$\sum_{f \in F_p} prvol_{pfst} = \sum_{s' \in S} dint_{pss't} + \sum_{c \in C} dext_{psct} + \sum_{f \in F_p} scupdem_{pfst} \qquad \forall p \in PI, s \in S, t \in T \quad (3.23)$$

Restriction (3.19) ensures that the distribution quantities do not exceed demand. It can be extended to include a minimum demand fill rate to enforce that certain markets are serviced even if this is not economically reasonable. Restriction (3.20) requires that the intermediate requirements induced by the production at a site as defined in equation (3.21) are met.

Restrictions (3.22) and (3.23) ascertain that production volumes at a site equal distribution quantities and quantities required for production trials in the context of product transfers. The restrictions also imply that no storage of products takes place. As separate variables are used for production and distribution quantities the model could however be extended to model inventory levels as well.

Capacity Constraints

$$\sum_{p \in P_f} \left(prvol_{pfst} \cdot capreq_{pst} \right) \leq tcap_{fst} \qquad \forall f \in F, s \in S, t \in T \quad (3.24)$$

$$tcap_{fst} = y_{fst} \cdot mtcap_{fs} + z_{fst} \cdot mtcap_{fs} \qquad \forall < f,s > \in FSA, t \in T \quad (3.25)$$

$$\underline{mtcap}_{fs} \cdot y_{fst} \leq tcap_{fst} \leq \overline{mtcap}_{fs} \cdot y_{fst} \qquad \forall < f,s > \in FSA, t \in T \quad (3.26)$$

$$\sum_{f \in F_G} y_{fst} \leq 1 \qquad \forall g \in G, s \in S, t \in T \quad (3.27)$$

$$\sum_{p \in P} \sum_{f \in F_p} prvol_{pfst} \cdot utreq_{pust} \leq utcap_{su} \qquad \forall < s,u > \in SU, t \in T \quad (3.28)$$

Equation (3.24) requires that the production volume allocated to a plant does not exceed technical capacity of the plant. Capacity consumption factors are site-specific to account for productivity differences and time dependent to allow for discrete process improvement projects. A plant's technical capacity is calculated by multiplying the number of production lines installed with the capacity per line (3.25). It can be varied between a lower and an upper bound (3.26). Restriction (3.27) states that from each plant class only one plant can be open at a site. The rationale behind this restriction is that only one plant type (size and degree of automation combination) should be put in place at a given site and capacity can be adjusted via the number of production lines. It is not implied that all production lines necessarily have to be in one building. For sites with utility capacity restrictions, constraint (3.28) ensures the available capacity is not exceeded.

Continuity of Capacity

$$y_{fst} \geq y_{fst-1} \qquad \forall < f,s > \in FSN, t \in T \setminus \{1\} \quad (3.29)$$

$$tcap_{fst} \geq tcap_{fst-1} \qquad \forall < f,s > \in FSN, t \in T \setminus \{1\} \quad (3.30)$$

$$z_{fst} - z_{fst-1} \leq w_{fst} \cdot BigM \qquad \forall < f,s > \in FSE, t \in T \setminus \{1\} \quad (3.31)$$

$$w_{fst} \geq w_{fst-1} \qquad \forall <f,s> \in FSE, t \in T \setminus \{1\} \quad (3.32)$$

$$y_{fst} \geq w_{fst} \qquad \forall <f,s> \in FSE, t \in T \setminus \{1\} \quad (3.33)$$

Restrictions (3.29) and (3.30) enforce that newly opened plants cannot be closed again during the planning horizon and the technical capacity installed at these plants cannot be reduced. Similarly, restrictions (3.31) to (3.33) enforce that for existing plants where capacity expansions took place throughout the planning horizon a plant closure is not possible. Production lines can however be closed based on the assumption that it might be desirable to shut down old equipment even if an expansion took place earlier.

Product-plant Allocation

$$\underline{S_p} \leq \sum_{s \in S} v_{pst} \leq \overline{S_p} \qquad \forall p \in P, t \in T \quad (3.34)$$

$$\sum_{f \in F_p} prvol_{pfst} \leq v_{pst} \cdot BigM \qquad \forall p \in P, s \in S, t \in T \quad (3.35)$$

$$v_{pst} \leq \sum_{f \in F_p} prvol_{pfst} \qquad \forall p \in P, s \in S, t \in T \quad (3.36)$$

Constraints (3.34) to (3.36) fix the number of sites that produce a product in parallel in any given time period between a lower and an upper bound. The intention is to be able to limit the network complexity while at the same time ensuring that supply of important products does not depend on a single production site. Restriction (3.36) is required to ensure that the "product is produced at site in time period" variable v is set to zero for sites where the product is not produced because no costs are associated with the allocation decision and hence no link to the objective function exists.

Product Transfers

$$prvol_{pfst} \leq u_{pfst} \cdot BigM \qquad \forall p \in P, <f,s> \in FSA, t \in T \quad (3.37)$$

$$u_{pfst} \geq u_{pfst-1} \qquad \forall p \in P, <f,s> \in FSA, t \in T \setminus \{1\} \quad (3.38)$$

$$u_{pfst} = 1 \qquad \forall <p,f,s> \in PFSE, t = 1 \quad (3.39)$$

$$scupdem_{pfst} \geq \sum_{p \in P} \sum_{f \in F} \left(u_{pfst} - u_{pfst-1}\right) \cdot scupvol_p \qquad \forall s \in S, t \in T \setminus \{1\} \quad (3.40)$$

$$scupdem_{pfst} \geq \sum_{p \in P} \sum_{f \in F} \left(u_{pfst} - scupstat_{pfs}\right) \cdot scupvol_p \qquad \forall s \in S, t = 1 \quad (3.41)$$

Before a product can be produced at a plant technically capable of doing so, a scale-up phase is required. Constraint (3.37) sets the scale-up variable to 1 if a product is produced at a plant while constraint (3.38) ensures the continuity of product-plant allocations. Equation (3.39) initializes the status at the beginning of the planning horizon and equations (3.40) and (3.41) create the secondary demand for production trials associated with a scale-up decision. The plant index is included to allow for a relaxation of restriction (3.27) which states that only one plant type per plant class can be established at a site.

Single-sourcing Restrictions

$$\sum_{s \in S} x_{pss't} \leq 1 \qquad \forall p \in PI, s' \in S, t \in T \quad (3.42)$$

$$dint_{pss't} \leq x_{pss't} \cdot BigM \qquad \forall p \in PI, s, s' \in S, t \in T \quad (3.43)$$

$$x_{pss't} \leq dint_{pss't} \qquad \forall p \in PI, s, s' \in S, t \in T \quad (3.44)$$

$$dext_{psct} \geq demand_{pct} \cdot minsupply_{psc} \qquad \forall < p, s, c > \in PSC^{+}, t \in T \quad (3.45)$$

$$dext_{psct} = 0 \qquad \forall < p, s, c > \in PSC^{-}, t \in T \quad (3.46)$$

Restrictions (3.42) to (3.46) implement the single-sourcing restrictions discussed in Chapter 3.3.6. The allocation of a dedicated intermediate source for every site using internally produced intermediates is achieved via equations (3.42) to (3.44).While the intention of this restriction is to ensure a homogenous quality of intermediates supplied, it is also exploited in the import duty calculations. Restriction (3.45) covers the case where for certain customers supply from a pre-specified site is to be ensured. Generally, dedicated plant-market allocations are not required in specialty chemicals but a model formulation analogous to the one for intermediates could be employed if required. Finally, restriction (3.46) blocks site-market allocations that are not permitted, e.g., because of trade restrictions.

Calculation of Net Tariff Costs

$$rtariff_{st} \geq \left[\begin{array}{l} \sum_{p \in P} \sum_{c \in CNE_s} \sum_{r \in R_p} \left[\begin{array}{l} dext_{psct} \cdot rreq_{prst} \cdot rpr_{rst} \\ \cdot \dfrac{exrate_{q(rs)t}}{exrate_{q(s)t}} \cdot tariff_{rc(rs)c(s)} \end{array} \right] \\ + \sum_{i \in I} \left[\begin{array}{l} \sum_{r \in R_i} \left[\sum_{s' \in SNE_s} dint_{iss't} - \sum_{p \in P} \sum_{c \in CE_s} dext_{psct} \cdot intreq_{pist} \right] \\ \cdot rreq_{irst} \cdot rpr_{rst} \dfrac{exrate_{q(rs)t}}{exrate_{q(s)t}} \cdot tariff_{rc(rs)c(s)} \end{array} \right] \end{array} \right] \quad \forall s \in S, t \in T \quad (3.47)$$

$$itariff_{is'st} \geq \left[\begin{array}{l} dint_{is'st} \\ - \sum_{p \in P} \sum_{c \in CE_s} dext_{psct} \cdot intreq_{pist} \\ - \sum_{p \in PI} \sum_{s'' \in SE_s} dint_{pss''t} \cdot intreq_{pist} \end{array} \right] \cdot \left[\begin{array}{l} trpice_{is't} \\ + \dfrac{trcost_{is'st}}{contvol_p} \end{array} \right] \cdot \frac{tariff_{ic(s')c(s)}}{exrate_{q(s)t}} \quad \begin{array}{l} \forall i \in PI \\ \forall s, s' \in S \\ \forall t \in T \end{array} \quad (3.48)$$

Equation (3.47) calculates the tariffs that have to be paid for import of raw materials. Duty drawbacks are considered by charging tariffs only for raw materials required for production of intermediate (second line) and finished goods (first line) that are not re-exported. The amount of intermediates not re-exported is adjusted for those intermediates that are transformed into goods subsequently re-exported at the same site. The formulation rests on the assumption that only one raw material source is used per site and that if a required intermediate is available locally the local source is used. Prices are converted from the currency used by the raw material supplier to the currency of the consuming site. Same as with the valuation of finished goods for tariff calculation the full transport costs are included in the tariff value of the raw materials and intermediates. Import tariffs for intermediates imported from another site are calculated in equation (3.48). The formulation differs from the one for raw materials because the source of the intermediates is not predetermined and the transfer price does not contain transportation costs. The net volumes of the intermediates which are not re-exported are calculated by subtracting the quantities contained in exported products from the total quantity imported. A non-negativity constraint sets the value to 0 if the respective intermediate is not imported from site s'.

It should be noted that the structure chosen here to model tariffs is based on assumptions specific to the problem instances analyzed (especially by exploiting the single sourcing assumption). Modifications might be required for example in order to allow drawbacks across a higher number of production levels, simultaneous supply from several suppliers or to accommodate activities such as toll processing and subsequent re-imports.

Personnel Requirements

$$facFTE_{fst} = \begin{bmatrix} fixFTE_{fs} \cdot y_{fst} \\ + \sum_{p \in P} prvol_{pfst} \cdot \frac{capreq_p}{mtcap_{fs}} \cdot varFTE_{fs} \end{bmatrix} \quad \forall < f,s > \in FSA, t \in T \quad (3.49)$$

Equation (3.49) calculates the variable personnel requirements associated with the production volumes assigned to a plant.

Initialization of Technical Capacity

$$y_{fst} = 0 \quad \forall < f,s > \in FSN, t = 1 \quad (3.50)$$

$$y_{fst} = 1 \quad \forall < f,s > \in FSE, t = 1 \quad (3.51)$$

$$z_{fst} = existmod_{fs} \quad \forall < f,s > \in FSE, t = 1 \quad (3.52)$$

Restriction (3.50) states that no new facilities can be opened in the first period of the planning horizon. Restrictions (3.51) and (3.52) initialize the technical capacity of existing facilities.

Non-negativity Constraints

$$capex_{st} \geq 0 \quad \forall s \in S, t \in T \quad (3.53)$$

$$capex_{fst} \geq 0 \quad \forall f \in F, s \in S, t \in T \quad (3.54)$$

$$cfix_{st} \geq 0 \quad \forall s \in S, t \in T \quad (3.55)$$

$$closurecost_{st} \geq 0 \quad \forall s \in S, t \in T \quad (3.56)$$

$$closurecost_{fst} \geq 0 \quad \forall f \in F, s \in S, t \in T \quad (3.57)$$

$$cvar_{st} \geq 0 \quad \forall s \in S, t \in T \quad (3.58)$$

$$dext_{psct} \geq 0 \quad \forall p \in P, s \in S, c \in C, t \in T \quad (3.59)$$

$$dint_{pss't} \geq 0 \quad \forall p \in P, s, s' \in S, t \in T \quad (3.60)$$

$$distrcost_{ct} \geq 0 \quad \forall c \in C, t \in T \quad (3.61)$$

$$facFTE_{fst} \geq 0 \qquad \forall f \in F, s \in S, t \in T \quad (3.62)$$

$$imptariff_{ct} \geq 0 \qquad \forall c \in C, t \in T \quad (3.63)$$

$$inventcc_t \geq 0 \qquad \forall t \in T \quad (3.64)$$

$$itariff_{is'st} \geq 0 \qquad \forall i \in PI, s, s' \in S, t \in T \quad (3.65)$$

$$netrev_{ct} \geq 0 \qquad \forall c \in C, t \in T \quad (3.66)$$

$$prvol_{pfst} \geq 0 \qquad \forall p \in P, f \in F, s \in S, t \in T \quad (3.67)$$

$$rtariff_{st} \geq 0 \qquad \forall s \in S, t \in T \quad (3.68)$$

$$scupcosts_{st} \geq 0 \qquad \forall s \in S, t \in T \quad (3.69)$$

$$scupdem_{pfst} \geq 0 \qquad \forall p \in P, f \in F, s \in S, t \in T \quad (3.70)$$

$$secdemand_{pst} \geq 0 \qquad \forall p \in PI, s \in S, t \in T \quad (3.71)$$

$$sevpay_{st} \geq 0 \qquad \forall s \in S, t \in T \quad (3.72)$$

$$shipcost_{st} \geq 0 \qquad \forall s \in S, t \in T \quad (3.73)$$

$$tcap_{fst} \geq 0 \qquad \forall f \in F, s \in S, t \in T \quad (3.74)$$

$$ttariff_{st} \geq 0 \qquad \forall s \in S, t \in T \quad (3.75)$$

Restrictions (3.53) to (3.75) are the common non-negativity constraints.

Variable Domains

$$u_{pfst} \in [0,1] \qquad \forall p \in P, < f, s > \in FSA, t \in T \quad (3.76)$$

$$v_{pst} \in [0,1] \qquad \forall p \in P, s \in S, t \in T \quad (3.77)$$

$$w_{fst} \in [0,1] \qquad \forall < f, s > \in FSE, t \in T \quad (3.78)$$

$$x_{pss't} \in [0,1] \qquad \forall p \in P, s, s' \in S, t \in T \quad (3.79)$$

$$y_{fst} \in [0,1] \qquad \forall <f,s> \in FSA, t \in T \quad (3.80)$$

$$z_{fst} \in [0,1,2,\ldots,n] \qquad \forall <f,s> \in FSA, t \in T \quad (3.81)$$

Constraints (3.76) to (3.81) are the common binary and integer constraints.

3.4.3 Model Extensions

Maximization of After-tax Cash Flows

In order to maximize the net present value of after-tax cash flows the model proposed in Chapter 3.4.2 has to be extended to determine the taxes payable in each country. To this end, pre-tax country profits comprising profits realized at both production and distribution entities have to be calculated. While the pre-tax profit of distribution entities can be calculated easily by subtracting all costs incurred from revenues realized, additional adjustments are required for production entities. Instead of cash flows associated with capital investments, depreciation costs have to be considered to identify pre-tax profits. The following assumptions are made to simplify the calculation:

- A single tax rate is assumed for every country (site-specific tax rates could be defined to accommodate cases where tax holidays are granted as a form of investment incentive).
- Accumulation of losses is not modeled. The effective tax rate can be set to zero for time periods where accumulated losses offset profits.
- Write-offs associated with plant/production line shutdowns are not considered.
- A straight-line depreciation with the depreciation period being at least as long as the planning horizon is assumed.
- All taxes are assumed to be payable at the country level based on profits realized across production and distribution entities.
- Special tax constructs such as a principal trading company are not modeled.

Table 9 contains the additional indices and parameters required to model tax effects:

Table 9. Additional indices and parameters for after-tax NPV

S_c	Sites located in country c
$COGS_{ct}$	Costs of goods sold in country c in period t
$deprcost_{st}$	Depreciation costs incurred at site s in period t
$deprexistass_{ct}$	Depreciation on existing assets at site s in period t
$deprr_c$	Depreciation rate in country c
$distrEBT_{ct}$	Pre-tax profits from product sales in country c in period t
EBT_{st}	Pre-tax profit at site s in period t
rev_{st}	Revenues realized at site s in period t
tax_{ct}	Taxes payable in country c in period t
$taxrate_{ct}$	Effective tax rate in country c in period t

$$distrEBT_{ct} = netrev_{ct} - imptariff_{ct} - distrcost_{ct} - COGS_{ct} \qquad \forall c \in C, t \in T \quad (3.82)$$

$$COGS_{ct} = \sum_{p \in P} \sum_{s \in S} dext_{psct} \cdot \frac{trprice_{pst}}{exrate_{q(c)t}} \qquad \forall c \in C, t \in T \quad (3.83)$$

Pre-tax profits from sales are calculated by subtracting all costs allocated to the selling country from net revenues. The costs of goods are calculated based on the transfer prices in equation (3.83).

$$EBT_{st} = \begin{bmatrix} rev_{st} \\ -\sum_{p \in PI} \sum_{s' \in S \setminus \{s\}} dint_{ps'st} \cdot \frac{trprice_{pt}}{exrate_{q(s)t}} \\ -cvar_{st} \\ -cfix_{st} \\ -ttariff_{st} \\ -deprcost_{st} \\ -shipcost_{st} \\ -scupcost_{st} \\ -closurecost_{st} \\ -sevpay_{st} \end{bmatrix} \qquad \forall s \in S, t \in T \quad (3.84)$$

$$rev_{st} = \begin{bmatrix} \sum_{p \in PE} \sum_{c \in C} dext_{psct} \cdot \frac{trprice_{pt}}{exrate_{q(s)t}} \\ + \sum_{p \in PI} \left[\sum_{s' \in S \setminus \{s\}} dint_{pss't} + \sum_{c \in C} dext_{psct} \right] \cdot \frac{trprice_{pt}}{exrate_{q(s)t}} \end{bmatrix} \qquad \forall s \in S, t \in T \quad (3.85)$$

$$deprcost_{st} = deprexistass_{st} + \sum_{t=1}^{t} capex_{st} \cdot deprr_c \qquad \forall s \in S, t \in T \quad (3.86)$$

Profits at the site level are calculated in equation (3.84) by subtracting all costs incurred at the site from the revenues realized at the site from finished and intermediate products as calculated by equation (3.85). In addition to the cost items already contained in the basic model formulation, the costs of intermediates received from other sites have to be allocated to the receiving site for tax purposes and the depreciation costs have to be subtracted instead of investment expenditures. Equation (3.86) defines that depreciation costs are calculated by adding to the depreciation plan of old assets (which has to be determined outside the model) depreciation incurred from new investments. Based on the assumption of straight-line depreciation and a depreciation period that exceeds the planning horizon these costs can be calculated as a fixed percentage of the total investments incurred at a site.

$$tax_{ct} = \begin{bmatrix} distrEBT_{ct} \\ + \sum_{s \in S_c} EBT_{st} \end{bmatrix} \cdot taxrate_{ct} \qquad \forall c \in C, t \in T \quad (3.87)$$

$$tax_{ct} \geq 0 \qquad \forall c \in C, t \in T \quad (3.88)$$

The resulting taxes payable per country are calculated in equation (3.87). This term has to be included in the objective function. Restriction (3.88) is a non-negativity constraint required to ensure that no negative taxes occur.

Explicit Modeling of Product Mix Complexity

As discussed in Chapter 3.3.4 product mix complexity effects can be modeled with a fixed cost charge incurred for every product allocated to a plant and/or via an estimation of the capacity lost due to more frequent/more complex changeover processes. Since strategic models such as the one proposed here do not simultaneously determine production campaigns, the latter approach requires an estimation of the number of changeovers occurring at a plant.

A feasible estimation approach can be developed analogous to the major/minor changeover concept Bitran and Gilbert (1990) use to model sequence-dependent setup costs. Products are grouped into changeover families with changeovers within a family not significantly exceeding regular cleaning requirements between two batches of the same product. Assuming that comprehensive cleanings required independent of changeover family switches are already deducted from each production line's technical capacity, a lower bound on the number of major changeovers required can be calculated by subtracting from the number of changeover families allocated to a plant the number of production lines available. Table 10 contains the additional indices, parameters and decision variables needed to implement both the fixed cost charge and the basic changeover family concept.

Table 10. Indices, parameters and decision variables for product mix extension

$h \in H$	Changeover families
P_h	Products belonging to changeover family h
$alloccost_{pt}$	Fixed costs for allocating product p to a plant in period t
$mixcost_{st}$	Product mix costs incurred at site s in period t
$xonumb_{fst}$	Minimum number of major changeovers required at plant f
$xofamall_{hfst}$	1, if changeover family h is allocated to plant f at site s in period t; 0 else

Additionally, the following modifications and extensions to the basic model are required:

$$mixcost_{st} = \sum_{f \in F} \sum_{p \in P_f} v_{pfst} \cdot \frac{alloccost_{pt}}{exrate_{q(s)t}} \qquad \forall s \in S, t \in T \quad (3.89)$$

If a fixed-cost charge to account for higher overhead costs, inventory requirements, etc. is to be charged for every product-plant allocation, equation (3.89) has to be added to the model and the resulting cost position has to be included in the objective function.

$$xofamall_{hfst} \cdot BigM \geq \sum_{p \in P_h} v_{pfst} \qquad \forall h \in H, <f,s> \in FSA, t \in T \quad (3.90)$$

$$xonumb_{fst} \geq \sum_{h \in H} xofamall_{hfst} - (z_{fst} + 1) \qquad \forall <f,s> \in FSA, t \in T \quad (3.91)$$

$$xonumb_{fst} \geq 0 \qquad \forall <f,s> \in FSA, t \in T \quad (3.92)$$

Restriction (3.90) sets the changeover family – plant allocation variable to 1 if at least one product from a changeover family is allocated to a plant. The lower bound on the number of major changeovers is calculated by subtracting from the number of changeover families allocated to a plant the number of production lines open at the plant (3.91). This approach assumes the entire production volume is produced in a single campaign. Additionally, an average campaign size could be considered if the assumption of one campaign per time period is unrealistic to account for additional major changeovers. A non-negativity constraint is added in (3.92) to account for cases where the number of production lines exceeds the number of changeover families.

Product mix complexity does not only reduce the technical capacity of a plant but during the time required for the changeovers personnel and possibly utilities such as steam or water are required. To include these effects in the model, a "changeover-product" can be defined with the product's capacity consumption factor and bill of materials specifying the resource consumption of a major changeover process. Restriction (3.93) converts the number of major changeovers into the corresponding demand for the changeover "product".

$$prvol_{pfst} = xonumb_{fst} \qquad \forall p = \text{"changeover"}, <f,s> \in FSA, t \in T \quad (3.93)$$

Make or Buy – Decisions and Vendor Selection

The basic model presented in Chapter 3.4.2 distinguishes between internally manufactured intermediates and externally procured raw materials without considering make or buy options for intermediates. For some application cases it might however be required to include make or buy – decisions in the network design model. The decision can be made either for the entire production network or individually for each site. In order to incorporate make or buy – decisions (and possibly vendor selection), suppliers have to be modeled as an additional network node. Table 11 contains the additional indices, parameters and decision variables required to implement a make or buy formulation for intermediates.

Table 11. Indices parameters and decision variables for make or buy extension

$v \in V$	Vendors
V_p	Vendors capable of supplying intermediate p
$pprice_{pvt}$	Ex works price of intermediate p from vendor v in period t
$vtariff_{ivst}$	Tariff costs incurred for intermediate i from vendor v at site s in period t
$svol_{pvst}$	Volume of intermediate p sourced from vendor v for site s in period t
xs_{pvst}	1, if intermediate p is supplied from vendor v to site s in period t; 0 else

$$xs_{pvst} \cdot BigM \le svol_{pvst} \qquad \forall p \in PI, v \in V_p, s \in S, t \in T \quad (3.94)$$

$$xs_{pvst} \le svol_{pvst} \qquad \forall p \in PI, v \in V_p, s \in S, t \in T \quad (3.95)$$

$$\sum_{s\in S} x_{pss't} + \sum_{v\in V} xs_{pvs't} \leq 1 \qquad \forall\, p \in PI, s' \in S, t \in T \quad (3.44a)$$

$$\sum_{s'\in S} dint_{ps'st} + \sum_{v\in V_p} svol_{pvst} = secdemand_{pst} \qquad \forall\, p \in PI,\, s \in S, t \in T \quad (3.20a)$$

To enforce the single sourcing restriction assumed for intermediates equation (3.94) sets the decision variable indicating that an intermediate is sourced from a certain vendor to 1 if a material flow between a vendor and a site exists for this material. Equation (3.44) from the basic model has to be modified to include the choice between internal and external sources for those intermediates where make or buy options exist. Flow constraint (3.20) has to be modified to ascertain that secondary demand is met either from internal or external sources.

$$cvar_{st} = \left[\begin{array}{l} \sum_{f\in F}\sum_{p\in P} prvol_{pfst} \left[\begin{array}{l} \sum_{r\in R_p} rreq_{prst} \cdot rpr_{rst} \cdot \dfrac{exrate_{q(rs)t}}{exrate_{q(s)t}} \\ + \dfrac{capreq_p}{mtcap_{fs}} \cdot varFTE_{fs} \cdot FTEcost_{st} \\ + othervarcost_{pst} \end{array} \right] \\ + \sum_{p\in PI}\sum_{v\in V_p} svol_{pvst} \cdot pprice_{pvt} \cdot \dfrac{exrate_{q(v)t}}{exrate_{q(s)t}} \end{array} \right] \qquad \forall\, s \in S, t \in T \quad (3.6a)$$

$$shipcost_{st} = \left[\begin{array}{l} \sum_{p\in PI}\sum_{s'\in S} \dfrac{dint_{ps'st}}{contvol_p} \cdot trcost_{c(s')c(s)t} \\ + \sum_{p\in PI}\sum_{v\in V_p} \dfrac{svol_{pvst}}{contvol_p} \cdot trcost_{c(v)c(s)t} \end{array} \right] \cdot \frac{1}{exrate_{q(s)t}} \qquad \forall\, s \in S, t \in T \quad (3.7a)$$

The cost incurred for external sourcing of intermediates have to be included in the relevant cost positions. To this end equation (3.6) has to be modified to include the purchasing price of intermediates sourced externally and equation (3.7) to include the transport costs from the vendor to the destination site.

$$vtariff_{ivst} \geq \left[\begin{array}{l} svol_{ivst} \\ -\sum_{p \in P} \sum_{c \in CE_s} \left[\begin{array}{l} dext_{psct} \\ \cdot intreq_{pist} \end{array} \right] \\ -\sum_{p \in PI} \sum_{s'' \in SE_s} \left[\begin{array}{l} dint_{pss''t} \\ \cdot intreq_{pist} \end{array} \right] \end{array} \right] \cdot \left[\begin{array}{l} \left[\begin{array}{l} pprice_{ivt} \\ \cdot exrate_{q(v)t} \end{array} \right] \\ + \dfrac{trcost_{c(v)c(s)t}}{contvol_p} \end{array} \right] \cdot \frac{tariff_{ic(v)c(s)}}{exrate_{q(s)t}} \qquad \begin{array}{l} \forall i \in PI, s \in S \\ \forall t \in T, v \in V \end{array} \quad (3.96)$$

$$vtariff_{ivst} \geq 0 \qquad \forall i \in PI, s \in S, t \in T, v \in V \quad (3.97)$$

$$ttariff_{st} = \sum_{p \in PI} \left[\sum_{s' \in S} itariff_{ps'st} + \sum_{v \in V_p} vtariff_{pvst} \right] + rtariff_{st} \qquad \forall s \in S, t \in T \quad (3.9a)$$

Restriction (3.96) calculates the net tariff load incurred for externally procured intermediates. The formulation is analogous to the one used for internally manufactured intermediates and rests on the single sourcing assumption. The basis for calculating the tariff is the sum of ex works purchasing price and transport costs. As the purchasing price is denominated in the currency of the vendor and transport costs are assumed to be denominated in the home currency, a conversion into the currency of the receiving site is required. Finally, equation (3.9) has to be modified to include the net tariffs from import of externally procured intermediates. It should be noted again that if the single-sourcing restriction is relaxed a different approach towards modeling tariffs has to be chosen. Since the single-sourcing assumption was found to be reasonable due to its impact on process stability, the modeling approach employed has been preferred over a more general statement.

Shared Resources and Temporary Equipment Shutdown

The nature of the value chain considered can require modifications to the basic technical capacity representation proposed in Chapter 3.3.3. In some cases the different product families belonging to a value chain differ with respect to their processing requirements insofar as some product families require additional process steps performed on equipment dedicated to this purpose. Typically, in such a situation two or more production lines share the equipments required for the additional process step. Therefore, the extension will be referred to as *shared assets* extension. An example of such a value chain can be found in the case study in Chapter 5.2.1. Two cases might have to be considered:

- Usage of the additional capacity needs to be considered in order to avoid infeasible product mix allocations but process costs can be ignored.
- The process step causes significant additional costs for capacity installation, personnel and/or maintenance (material and utility costs are part of the recipe and are included irrespective of the capacity modeling approach selected) of the equipment and hence an explicit modeling of resource requirements is required in addition to capacity constraints.

Furthermore, the link of the shared asset capacity to the overall plant capacity has to be determined. If the equipment capacity is to be selected by the optimization model, an additional integer decision variable is required. Alternatively, a fixed capacity or a capacity correlated with overall plant capacity (e.g., 50% of total plant capacity or one unit for every two production lines) can be assumed. Additional restrictions are required if the correlation approach is to be combined with the modular capacity concept. The choice of the most appropriate variant depends on the characteristics of the equipment considered and the effects on overall model complexity.

Additionally, temporary shutdowns of production lines might be an option to reduce fixed costs in reaction to temporary demand fluctuations or to adjust production capacity to exchange rate-driven volume shifts. In order to retain the option to reactivate production lines at a later point in time limited maintenance will be performed to keep the equipment operational. Model formulations to integrate these aspects are provided below using the additional notation shown in Table 12.

Table 12. Indices, parameters and decision variables for shared resources extension

$e \in E$	Equipments
$capreq_{ep}$	Capacity of equipment e required per unit of product p
$ecap_{ef}$	Capacity per module of equipment e at facility f
$efcor_{efs}$	Correlation factor for capacity of equipment e at facility f at site s
$eFTE_{efs}$	Number of operators required per unit of equipment e at facility f at site s
$etcap_{efst}$	Capacity of equipment e available at facility f at site s in period t
$linemaint_{fs}$	% of fixed costs to maintain shut-down line at plant f at site s
$unitcapex_{efst}$	Capital expenditure per unit of equipment e at facility f at site s in period t
$unitfixcost_{efst}$	Fix operating costs per unit of equipment e at facility f at site s in period t
ze_{efst}	Number of equipments of type e at facility f at site s in period t
zf_{fst}	Number of prod. lines shut down at plant f at site s in period t

Capacity Restrictions

$$\sum_{p \in P_e} prvol_{pfst} \cdot capreq_{ep} \leq ze_{efst} \cdot ecap_{ef} \qquad \forall e \in E, <f,s> \in FSA, t \in T \quad (3.98)$$

$$(y_{fst} + z_{fst}) \cdot efcor_{efs} \leq ze_{efst} < (y_{fst} + z_{fst} + 1) \cdot efcor_{efs} \qquad \forall e \in E, <f,s> \in FSA, t \in T \quad (3.99)$$

$$ze_{efst} \in \{0,1,2,\ldots,N\} \qquad \forall e \in E, <f,s> \in FSA, t \in T \quad (3.100)$$

$$tcap_{fst} = y_{fst} \cdot mtcap_{fs} + (z_{fst} - zf_{fst}) \cdot mtcap_{fs} \qquad \forall <f,s> \in FSA, t \in T \quad (3.25a)$$

The additional capacity restriction (3.98) accounts for the capacity of the shared resource. In order to determine shared-resource capacity, restriction (3.99) can be used if the number of equipment units is correlated with the number of production lines installed at a plant. In combination with the integrality restriction (3.100) it enforces the step-wise increase of the shard resource capacity in line with the development of overall plant capacity. For example, if for every three production lines one equipment unit is to be installed, the second unit will be installed once the fourth production line is put into operation. If the model is to select the number of equipment units independently, restriction (3.99) has to be deactivated. Finally, for the option to temporarily shut down production lines capacity constraint 24 needs to be modified as shown above.

In case the costs for operating the equipment and/or capital expenditures to set up the equipment have to be considered explicitly, the variable (3.6) and fixed (3.8) cost calculations and the calculation of investment expenditures (3.11) have to be extended as shown below. Additionally, the fixed-cost reduction achieved by temporarily shutting down a production line has to be accounted for as shown in equation (3.8a).

Cost Effects

$$cvar_{st} = \sum_{f \in F} \sum_{p \in P} prvol_{pfst} \left[\begin{array}{l} \sum_{r \in R_p} rreq_{prst} \cdot rpr_{rst} \cdot \dfrac{exrate_{q(rs)t}}{exrate_{q(s)t}} \\ + \dfrac{capreq_p}{mtcap_{fs}} \cdot varFTE_{fs} \cdot FTEcost_{st} \\ + \dfrac{capreq_{ep}}{ecap_{efs}} \cdot eFTE_{efs} \cdot FTEcost_{st} \\ + othervarcost_{pst} \end{array} \right] \qquad \forall s \in S, t \in T \quad (3.6b)$$

$$cfix_{st} = \sum_{f \in F} \begin{bmatrix} y_{fst} \cdot \begin{bmatrix} fixcost_{fst} \\ + fixFTE_{fs} \cdot FTEcost_{st} \end{bmatrix} \\ + z_{fst} \cdot linefixcost_{fst} \\ - zf_{fst} \cdot linefixcost_{fst} \cdot (1 - linemaint_{fs}) \\ + ze_{efst} \cdot unitfixcost_{efst} \end{bmatrix} \qquad \forall s \in S, t \in T \quad (3.8a)$$

$$capex_{fst} \geq \begin{bmatrix} (y_{fst+1} - y_{fst}) \cdot plantcapex_{fst} \\ + (z_{fst+1} - z_{fst}) \cdot linecapex_{fst} \\ + \sum_{e \in E_f} (ze_{efst+1} - ze_{fst}) \cdot unitcapex_{efst} \end{bmatrix} \qquad \forall < f, s > \in FSA, t \in T \quad (3.11a)$$

The integration of capital expenditures for the shared resources as shown in equation (3.11a) rests on the assumption that the number of equipments is correlated to the number of production lines. If the model can independently select the shared resource capacity, it is theoretically possible that the number of shared resources operated increases while the number of production lines used decreases. In this case a separate calculation of the investment expenditures for shared resources is required to avoid that "negative" capital expenditures from capacity reductions that are eliminated via the non-negativity constraint (3.54) offset the expenditures for shared resource installations.

3.4.4 Accounting for Uncertainty: Robust Production Network Design

Despite the fact that stochastic optimization is often not applicable in practice, the uncertainty inherent in many of the external parameters has to be considered in the course of a production network analysis. Sensitivity and especially scenario analyses in the form of "what-if" – explorations of the solution space offer well-accepted approaches to explore the effects of parameter uncertainty on production network design (refer also to the more detailed discussion in Chapter 5.4). Typically, as described for example by Eppen et al. (1989, p. 519) for General Motors, in a scenario-based planning approach alternative scenarios are generated and the optimization model is solved for each scenario. The resulting network configuration alternatives are compared to get an understanding of the solution stability. The economic performance of the preferred alternative, usually the optimal solution obtained for the most likely scenario, can be further evaluated under the other scenarios by solving the optimization model with structural decision variables being fixed to this solution. If it performs badly under other relevant scenarios, more robust alternative solutions have to be found. However, these step-by-step analyses cannot ensure that all interdependencies are considered and for complex networks it may be difficult to find robust design alternatives. While a discussion of the various methods that can be used for optimization under uncertainty is not within the focus of this work, a robustness approach that can provide valuable insight is described below. For a review of the state-of-the-art in optimization under uncertainty and further references the reader is referred to Sahinidis (2004).

Two-stage Decisions with Recourse

In stochastic optimization a class of problems, already described by Dantzig (1955), exists that is referred to as two-stage decisions with recourse.[35] In a first stage, *design variables* such as capacity investment decisions have to be taken for a certain planning horizon without knowing the realization of uncertain parameters such as demands, prices or exchange rates. In a second stage, once the uncertain parameters materialize, *control variables* such as production volumes, product allocations, etc. have to be taken within the boundaries imposed by the structural decisions taken in the first stage (cf. Birge and Louveaux 1997, pp. 52-59). Obviously, production network design and capacity planning problems are typical examples of this problem class. The classical objective of two-stage recourse problems is to optimize the sum of first-stage design costs and the expected (mean) value of the uncertain second-stage recourse costs/cash flows (cf. Birge and Louveaux 1997, p. 54). Alternatively, in worst-case analyses the maximum deviation might be considered instead of the expected value (cf. Mulvey et al. 1995, p. 264). Mathematically, the models are solved for a finite but large number of discrete scenarios (which can be derived from sampling the underlying probability distribution). The resulting model formulations are also referred to as deterministic equivalent problem. Techniques to generate scenarios via discretization of random variables and solution approaches using Xpress SP are described in Dormer et al. (2005). Applications of this type of model for location/capacity planning have for example been reported in Bienstock and Shapiro (1988) or Wallace (1988).

[35] The problem might also entail multiple stages if it is intended to model multiple design decision periods. See Birge and Louveaux (1997, pp. 59-60) for a discussion of the issue.

Robust Optimization and Optimization with Restricted Recourse

In practice decision makers typically are risk averse and the expected value approach does not take into account the variability of the solutions obtained under the probability distributions or scenarios considered for the uncertain parameters. Rosenhead et al. (1972) introduced the aspect of *robustness* as a criterion for strategic planning to address this issue. Building on the notion of robustness, Mulvey et al. (1995) developed the concept of *robust optimization* distinguishing between two different types of robust models. A model is *solution robust* if the solution obtained remains close to optimality for any realization of the uncertain parameters. The model itself is robust if it remains (almost) feasible for any realization of the uncertain parameters (*model robust*).[36] Here, only solution robustness is of interest as the most important elements of uncertainty in production network design, namely demand volumes, costs, prices and exchange rates, should not lead to infeasibility problems under different scenarios considered.

The key distinction between an optimization of the expected profits/cost and the robust approach is that the former obtains a solution that possesses least expected costs but might cause extreme swings in the control variables and the economic performance whereas the robust model leads to less extreme swings but often possesses higher expected costs (cf. Mulvey et al. 1995, p. 270). In the robust optimization framework the objective function is extended to include a measure of the variability of the second-stage control variables. Hence, the approach can effectively be characterized as "a goal programming approach to balance the tradeoffs between expectation and variability of the recourse costs" (Ahmed and Sahinidis 1998, p. 1885). Mulvey et al. suggest to use either the mean/variance approach originally proposed by Markovitz (1959) for portfolio selection or the expected utility models proposed by von Neumann and Morgenstern (1953). In a similar approach, developed independently by Paraskevopoulos et al. (1991) for capacity planning in plastics industry, a penalty factor on the sensitivity of the second-stage costs to various types of uncertainty is included in the objective function. Gutierrez and Kouvelis (1995) apply the robust optimization approach to international sourcing in light of uncertain exchange rates. They develop an algorithm based on a minmax regret criterion to identify a number of best robust solutions. An overview of

---

[36] Assuming that it is often not possible to obtain a feasible solution under all possible realizations of the uncertain parameters, Mulvey et al. use a multicriteria objective function that penalizes infeasibilities to trade off model robustness and solution robustness.

different measures of robustness and a supply chain model based on the two-stage recourse approach can also be found in Mo and Harrison (2005).

An alternative approach to enforce robustness via additional constraints was proposed by Vladimirou and Zenios (1997). They include satisficing constraints on the recourse variability using the euclidian deviation of the second-stage solution vector from the expected solution. An iterative procedure is used to gradually enforce the recourse constraint (cf. Vladimirou and Zenios 1997, pp. 181-184). Alternatively, a constraint on the variability of the recourse costs could be used (cf. Ahmed and Sahinidis 1998, p. 1885). The advantages of the restricted recourse approach are among others that it remains possible to economically interpret the result of the objective function and that by gradually tightening the level of variability accepted, expected objective value/degree of robustness trade-offs can easily be analyzed.

As Ahmed and Sahinidis (1998, p. 1886) and Eppen et al. (1989, p. 523) point out, mean-variance robustness approaches lead to non-linear models that make problem instances of realistic size hard to solve. Furthermore, variance is a symmetric risk measure whereas in economic problems an asymmetric measure that only penalizes increased costs or reduced profits is more appropriate. Therefore, Eppen et al. (1989, p. 523) suggest to use a restriction that limits the Expected Downside Risk (EDR). EDR is defined as the expected value of profits that are below (costs that are above) a previously defined minimum value. By calculating the EDR of the unconstrained solution and gradually tightening the additional constraint on EDR it is possible to construct solutions that optimize the model subject to a tighter limit on expected downside risk. Further developing the EDR approach, Ahmed and Sahinidis (1998, pp. 1886-1887) suggest to use the Upper Partial Mean (UPM) instead to control variability. UPM is defined as the expectation of positive deviations over all scenarios considered. The key advantage of this variability measure, which can be used both for robust optimization and restricted recourse optimization, is that it does not require an a priori specification of the desired target level while it at the same time maintains the linearity of the resulting model. Ahmed and Sahinidis (1998) report findings from an application example to process planning in chemical industry where they demonstrate that for the case considered a reduction of the UPM by more than 40% can be achieved with an adverse effect of less than 1% on expected overall performance.

An Illustrative Example

Implementing the restricted-recourse approach requires an almost complete reformulation of the above-introduced optimization model. Since providing this reformulation would considerably inflate this chapter and the complexity of the model would make it hard to demonstrate the effects of the restricted recourse approach, a simple illustrative model is used instead. The exemplary model used below describes a planning problem where a single product is to be supplied to a single market from one ore more of a set of prospective facilities. Due to setup lead times all investment decisions have to be taken before the realization of the demand scenarios becomes known. The following deterministic optimization model could be used to solve the planning problem:

Table 13. Indices, parameters and decision variables for illustrative example

Index sets			
$f \in F$	Plants	$t \in T$	Time periods
Parameters			
cap_f	Capacity of facility f		
$cfix_f$	Fixed costs per period for operating facility f		
$cvar_f$	Unit production and transportation costs at facility f		
dem_t	Demand in period t		
$invest_f$	Investment required to set up facility f		
$pcap_f$	Production capacity of facility f		
$price_t$	Unit price in period t		
Decision variables			
x_{ft}	1, if facility f is open in period t; 0 else		
y_{ft}	Quantity supplied by facility f in period t		

Objective Function

$$max \sum_{t \in T} \sum_{f \in F} \left[-inv_f \cdot (x_{ft} - x_{ft-1}) - cfix_f \cdot x_{ft} + y_{ft} \cdot (price_t - cvar_f) \right] \qquad (3.101)$$

Subject to

$$\sum_{f \in F} y_{ft} \leq dem_f \qquad \forall t \in T \quad (3.102)$$

$$y_{ft} \leq pcap_f \qquad \forall f \in F, t \in T \quad (3.103)$$

$$x_{ft} \geq x_{ft-1} \qquad \forall f \in F, t \in T \quad (3.104)$$

$$x_{ft} \in [0,1] \qquad \forall f \in F, t \in T \quad (3.105)$$

$$y_{ft} \geq 0 \qquad \forall f \in F, t \in T \quad (3.106)$$

Table 14 contains an exemplary data set describing three potential production facilities. Additionally, Table 15 contains three different demand scenarios for the example.

Table 14. Facility data for illustrative example

Site	Setup costs	Fixed costs	Capacity	Unit costs
A	300	60	150	3
B	210	42	130	7
C	200	40	105	7

Table 15. Demand data for illustrative example

		Demand		Price	
Scenario	Probability	Period 1	Period 2	Period 1	Period 2
1	0.3	100	200	14	15
2	0.5	150	250		
3	0.2	175	320		

In a deterministic planning environment the most likely scenario, here scenario 2, would be considered the base case and the optimization model would be solved based on this scenario. The optimal decision would be to open facility 1 in period 1 and facility 3 in period 2 leading to a total profit of 2,590. To assess the robustness of this network to alternative demand scenarios the profit achievable with this configuration in case of the alternative demand scenarios can be assessed. In the example, for scenario 1 the overall profit would be 1,640 and for scenario 3, 2,765 respectively. Considered individually, the optimal decision for scenario 1 would be to open only facility 1 with a total profit of 1,880 and for scenario 3 to open both facilities 1 and 2 in period 1 with a total profit of 2,931. In order to explicitly incorporate the uncertainties caused by the different realization probabilities of the three demand scenarios, the optimization model can be extended into a two-stage decision with recourse:

Table 16. New indices and modified parameters/variables for restricted recourse

Index sets	
$s \in S$	Scenarios
Parameters	
dem_t^s	Demand in period t under scenario s
p^s	Probability of scenario s
Decision variables	
y_{ft}^s	Quantity supplied by facility f in period t under scenario s

Modified Objective Function

$$max \sum_{t \in T} \sum_{f \in F} \left[-inv_f \cdot (x_{ft} - x_{ft-1}) - cfix_f \cdot x_{ft} + \sum_{s \in S} p^s \cdot y_{ft}^s \cdot (price_t - cvar_f) \right] \qquad (3.107)$$

Using the probabilities provided in Table 15, a maximization of the expected profit would lead to the decision to open facility 1 in period 1 and facility 2 in period 2 with an expected profit of 2,321. Obviously, this network is not optimal for any of the given scenarios. Since maximizing the expected profit does not guarantee acceptable performance levels (i.e. robustness) if another scenario materializes, the recourse formulation is further extended to gradually reduce the variability of the solution obtained via the restricted recourse approach using the partial mean variability measure proposed by Ahmed and Sahinidis (1998). To this end the following additions are required:

Table 17. Additional parameters for Upper Partial Mean restriction

Parameters	
ND^s	Negative deviation of scenario s from expected performance
ε	Maximum partial mean deviation allowed

Additional Restrictions

$$ND^s \geq \left[\begin{array}{l} -\sum_{t \in T} \sum_{f \in F} y_{ft}^s \cdot (price_t - cvar_f) \\ +\sum_{t \in T} \sum_{f \in F} \sum_{s \in S} p^s \cdot y_{ft}^s \cdot (price_t - cvar_f) \end{array} \right] \qquad \forall s \in S \; (3.108)$$

$$ND^s \geq 0 \qquad \forall s \in S \; (3.109)$$

$$\sum_{s \in S} p^s \cdot ND^s \leq \varepsilon \tag{3.110}$$

The combination of restriction (3.108) with the non-negativity condition in restriction (3.109) determines for each scenario the amount by which the profit obtainable if the respective scenario materializes falls below the expected profit. The sum of this deviation over all scenarios as expressed by equation (3.110) is the expected negative deviation from the mean profit or negative partial mean. By gradually restricting the maximum value of the partial mean deviation alternative network configurations with an increased level of robustness can be identified. As already discussed above the increase in robustness is associated with a reduction of the expected profit. In the case example an iterative restriction of the partial mean would first lead to a shift from opening plant 2 in period 2 to opening plant 3 in period 2 and then to a network relying only on plant 1. Table 18 summarizes the characteristics of the solutions obtained.

Table 18. Robust network design alternatives

Scenario	Expected profit	Expected partial mean	Profit in case of Scenario 1	Scenario 2	Scenario 3
• Site A in period 1, Site B in period 2	2,321	213.9	1,608	2,558	2,798
• Site A in period 1, Site C in period 2	2,313	202.2	1,640	2,590	2,630
• Site A in period 1	2,265	115.8	1,880	2,430	2,430

Applicability to Production Network Design

From a mathematical perspective the two-stage optimization with restricted recourse approach presented above can be considered a deterministic equivalent formulation of a stochastic optimization problem (cf. Birge and Louveaux 1997, p. 84). Employing the restricted recourse framework in the sense of stochastic optimization requires to establish probability distributions for all uncertain parameters and derive a (sufficiently large) number of discrete scenarios from these distributions. To derive these discrete scenarios techniques such as Monte Carlo sampling, for example described in Birge and Louveaux (1997, pp. 331-352), can be used. However, in practice two types of problems typically render this approach infeasible. On the one hand model size grows multiplicatively with the number of scenarios included in the analysis thereby drastically increasing model complexity (cf. Bienstock and Shapiro 1988, p. 217). On the other hand the main reason for initially ruling out a stochastic optimization approach was that the difficulties associated with the specification of probability distributions for the relevant parameters severely inhibit the practical applicability of stochastic models in strategic planning.

Nevertheless, the restricted recourse approach can be a valuable tool in a scenario-based planning approach. Discussing the two-stage stochastic programming approach, Bienstock and Shapiro (1988) use a model for capacity planning in electrical utilities to demonstrate its usefulness. The example they use considers alternative scenarios for two parameters, namely demand and environmental regulations. The authors combine three demand scenarios and two environmental scenarios. Based on the resulting six scenarios they show how a capacity plan hedged against uncertainties can be developed. Similarly, Eppen et al. (1989) show that the restricted recourse approach can be used to identify more robust solutions in an environment where, as is often the norm in strategic planning, only a limited number of expert-defined scenarios are available.

An important aspect of employing the robustness approach is to properly select the time periods and design elements that require a decision in the first stage before any additional information on the uncertain parameters is available. If for example a ten-year planning horizon is considered, clearly not all location or capacity modification decisions belong to the first stage. However, with an increasing number of structural decisions (integer variables) in the second stage, model complexity increases even more because separate variables are required for every scenario.

3.5 Numerical Performance

Numerical tests with different data sets derived from real data collected at the industrial partner in the course of the pilot application reported in Chapter 5 were performed to establish the applicability of the proposed model to problem instances of realistic size. While the numerical performance of mixed-integer programs depends to a large extent on the data set used, some results are provided below. All tests were performed using ILOG OPL 4.2 and CPLEX 10 on a computer with an AMD Athlon XP 2600+ processor and 1 GB memory using a ten-year planning horizon.

In a first step data sets with increasing complexity with respect to the number of products, markets and potential production facilities were generated. The results of the numerical tests are shown in Table 19. Product and market complexity was increased by further disaggregating the demand without changing the overall demand volume. As was expected the calculation times (all calculations were performed with a MIP gap of 0%) increased with the disaggregation of the demand data. Based on the data set with the greatest level of detail variations were created by assuming different growth rates for volumes and prices throughout the planning horizon. As can be seen calculation times varied considerably despite an identical number of constraints, parameters and variables. Similarly, a change of the exchange rates used in the model led to a considerable change in calculation times (calculation run 3 versus 7). Finally, based on the largest data set and the stable demand situation the number of potential facilities was extended from 5 to 16. As can be seen this drastically increased the calculation time to 18 hours.

Table 19. Effect of network complexity on calculation times

Run	Product/market combinations	Data set characteristics	Plants Existing	Plants Potential	MIP-GAP	CPU time (hh:mm:ss)
1	41	stable	9	5	0	00:02:52
2	77	stable	9	5	0	00:04:51
3	151	stable	9	5	0	00:06:18
4	151	volume + 5% p.a.	9	5	0	05:27:57
5	151	volume - 5% p.a.	9	5	0	00:10:40
6	151	volume +5%, price +3% p.a.	9	5	0	04:08:18
7	151	stable, other exchange rates	9	5	0	00:18:15
8	151	stable	9	16	0	18:29:16

While calculation times are not as critical in strategic planning applications as in operational environments, the ability to perform scenario analyses and interactively explore the solution space require relatively short calculation times. Based on a data set built from the stable demand data used above with random volume and price growth, strategies to reduce calculation time were tested (cf. Table 20). A significant reduction of the calculation time was achieved by deactivating the restriction on the number of sites a product can be allocated to (equation (3.34) on page 101). This is probably due to the fact that in the basic model no cost effects are associated with product-plant allocations and hence the BigM-restriction is not linked to the objective function.

In a second step the effect of using different MIP gaps was tested. Since the data underlying the calculations is based on highly uncertain forecasts a MIP gap of 0% is not required. As can be seen with a MIP gap of around 1% calculation times of less than 3 minutes can be achieved. This is clearly sufficient to interactively evaluate a broad number of scenarios. To understand the effects different MIP gaps have on the solutions obtained a comparison of the resulting network designs was conducted. Typically, the sites and capacities selected were identical across different MIP gaps but the time period the decisions were implemented varied. If the decision makers desire a solution with a gap of 0% a practical approach is to perform a final calculation run without allowing a MIP gap after all interactive analyses have been performed.

Table 20. Reduction of calculation times

Run	Product/market combinations	Data set characteristics	Plants Existing	Plants Potential	MIP-GAP	CPU time (hh:mm:ss)
1	151	-	9	16	0	31:31:18
2	151	deactivated prod-plant alloc.	9	16	0	05:16:42
3	151	deactivated prod-plant alloc.	9	16	5%	00:02:54
4	151	deactivated prod-plant alloc.	9	16	1%	00:02:04
5	151	deactivated prod-plant alloc	9	16	0,5%	00:13:21
6	151	deactivated prod-plant alloc.	9	16	0,1%	02:57:21

4 Evaluation of Individual Production Sites

The need to evaluate production sites can arise in different situations. Most obviously, a network optimization project resulting in the decision to establish new production capacity may require a site selection phase in one or more countries (cf. Chap. 2.4.4). Depending on the status quo, the task either is to identify and evaluate potential production sites or to choose the most suitable one from a set of existing sites. Conversely, if capacity is to be reduced potential closure candidates might have to be assessed to identify the one least suitable for future use. Additionally, as pointed out in Chapter 2.4.5, a regular evaluation of all production sites is also required in the context of site controlling. Here the objective is to rank a company's entire portfolio of existing sites to identify action needs.

As discussed in Chapter 2.2.2 a broad range of criteria have to be considered if an in-depth assessment of individual sites is required. In practice matters are further complicated by the fact that the majority of sites host plants from multiple value chains. In this chapter a uniform decision support tool is developed to ensure consistent evaluations in all instances requiring site assessments. To this end Chapter 4.1 introduces the field of Multiple Criteria Decision Analysis (MCDA). Two different families of tools that could be applied to the decision problem at hand are discussed in greater detail in Chapters 4.2 and 4.3 respectively. As the use of Data Envelopment Analysis (DEA) for multiple criteria decision problems has been proposed in literature, too, the method is introduced in Chapter 4.4. An evaluation model for specialty chemicals production sites developed in cooperation with the industrial partner is presented in Chapter 4.5 and insights from application case studies are reported.

4.1 Introduction to Multiple Criteria Decision Analysis

4.1.1 Classification of MCDA Methods

The number and diversity of sometimes conflicting location factors that have to be considered when selecting/ranking production sites make obvious the need to employ a systematic decision support tool. Research in the field of Multiple Criteria Decision Analysis (MCDA)[37] aims at providing tools that guide decision makers in identifying their most preferred solution to such a decision problem (cf. Stewart 1992, p. 571). Hanne (1998, p. 11) estimates that at least 100 different MCDA methods have been published in 3,000 – 5,000 publications. The reader interested in a comprehensive compilation on the subject can refer to the collection of state of the art surveys edited by Figueira et al. (2005). It contains a chapter on each major MCDA method and recent research trends. Additionally, the books by Belton and Stewart (2002) and Vincke (1992) are excellent starting points for those whishing to acquaint themselves with the subject and a manual prepared by Dodgson et al. (2000) for use within the British government is available online[38]. A brief outline of the major principles underlying decision analysis and their relevance for applications in industry is provided by Phillips (1986, pp. 190-193).

However, only a subset of the methods available is useful for the decision problem at hand. It is common to differentiate between two types of multiple criteria decision situations (cf. Hwang and Yoon 1981, pp. 2-4; Mustafa and Goh 1996, pp. 169-170):

- *Multiple Attribute Decision Analysis (MADA)* supports either selecting the most preferred alternative from an explicitly specified set of alternatives or ranking/clustering of all alternatives within this set.
- *Multiple Objective Decision Analysis (MODA)* designs the most preferred alternative within a (usually continuous) solution space using a mathematical programming structure to optimize the level of a set of quantifiable objectives.

---

[37] Some authors also use the term "Decision Aid" synonymously. In the past, also the terms "Decision Making" and "Decision Science" were used. However, as discussed by Roy (1993) and Roy (1990a), the latter terms imply a level of objectivity that cannot be achieved in real-life decision situations.

[38] http://www.communities.gov.uk/pub/252/Multicriteriaanalysismanu-alPDF1380Kb_id1142252.pdf (as of 09/30/2006)

The assessment/ranking of potential or existing production sites clearly falls into the realms of MADA. Consequently, this work does not consider MODA tools. For a discussion of MODA methods and further references the reader is referred to Steuer (1986) and Hwang and Masud (1979).

Hwang and Yoon (1981, pp. 8-9) developed the classification of MADA methods shown in Figure 24. It is based on the type and quality of information available on decision makers' preferences. The tools most suitable for site assessments belong to the subset of tools that work with cardinal preference information. Major methods from within this subset will be introduced below while the reader is referred to Hwang and Yoon (1981) for the other types of MADA tools.

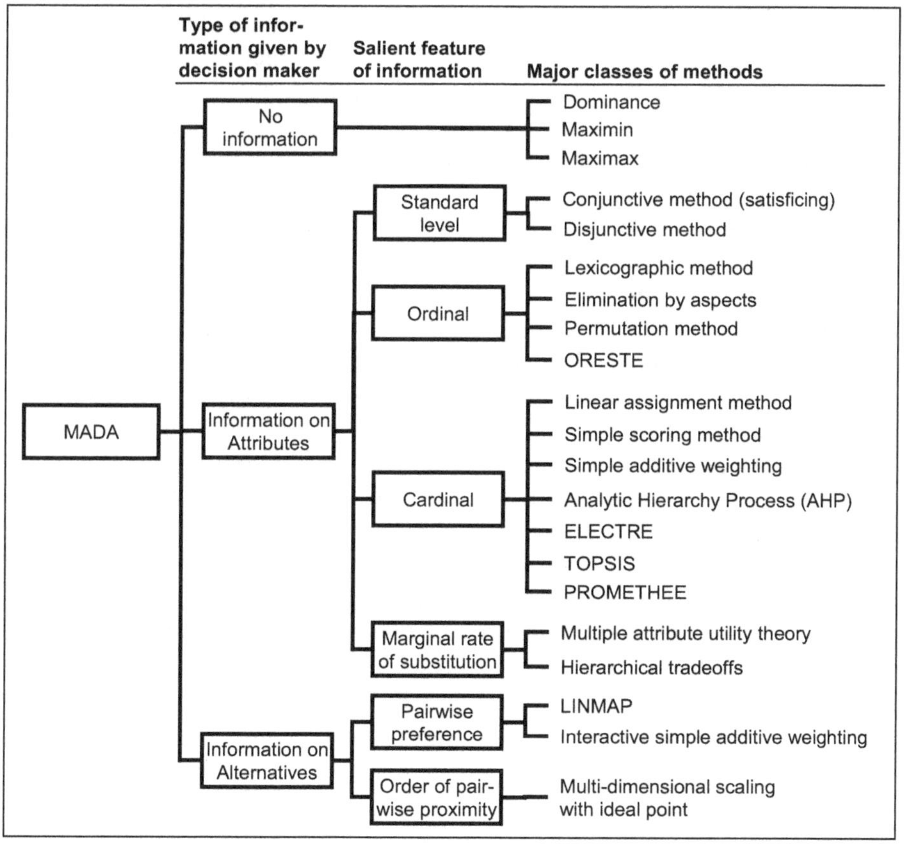

Fig. 24. Overview multiple attribute decision analysis tools[39]

[39] Source: Hwang and Yoon (1981, p. 8) with modifications by the author.

Two different philosophies of how to arrive at a synthesized recommendation exist (cf. Vincke 1992, p. xvi). The traditional, American school seeks to arrive at a single synthesizing value indicating the decision maker's preference assuming a fully compensatory model. The so-called technological or French approach builds an outranking relationship between alternatives that measures concordance and discordance separately and allows for incomparability. A separate synthesizing step may follow to arrive at an aggregated recommendation (cf. Roy 2005, p. 16). The main representatives of the first group are simple scoring, simple additive weighting, Analytic Hierarchy Process (AHP) and Multi-Attribute Value/Utility Theory (MAVT/MAUT) while the most common methods in the second group are the PROMETHEE and ELECTRE methods. The difference between MAVT and MAUT is that MAUT, extending classical decision theory, adds the use of probabilities and expectations to the methodology to deal with uncertainty. The utility function elicitation process to build a MAUT model is very complex. As deterministic models in combination with sensitivity analyses and/or scenarios in most cases can provide similar insights at strongly reduced complexity (cf. Belton and Stewart 2002, p. 95), MAUT will not be covered here. For a detailed discussion the reader should refer to Keeney and Raiffa (1976).

There is no consistent terminology within MADA research. In the remainder the terminology proposed by Keeney and Raiffa (1976, pp. 32-34) will be used. Decision makers typically seek to achieve an overall, broad *objective* (often also referred to as *goal*) such as selecting the *most suitable* site. Roy (1996, pp. 57-73) distinguishes between three *problematics* to describe different types of these broad objectives. In addition to the above-mentioned choice problematic, in the sorting problematic the decision maker intends to group alternatives into several categories while in the ranking problematic he requires a partial or complete preference order of all alternatives. While not being specific enough to directly evaluate alternatives, the overall objective serves as the starting point for defining operational lower-level *sub-objectives* (often referred to as "decision criteria"). At the lowest level *attributes* indicate to which extent the alternatives available meet the respective (sub-) objective.

4.1.2 Common Steps of MADA Methods

All MADA methods require a definition and structuring of objectives, attributes and alternatives and the majority of methods also requires a definition of weights that reflect the importance of different sub-objectives. In order to avoid repetitions in the description of the various MADA methods these two aspects are discussed below.

Definition and Structuring of Objectives, Attributes and Alternatives

Structuring a decision problem is an iterative process that ideally involves all major stakeholders and might lead to considerable revisions of the original problem formulation (cf. Roy 1999, pp. 1-6). In practice, the process in itself often is at least as valuable as the recommendation arrived at by later use of a decision analysis model (cf. Belton and Hodgkin 1999, p. 248; Goodwin and Wright 1998, p. 397; Phillips 1986, p. 189).

Different approaches can be used to identify the objectives in a decision problem. Keeney and Raiffa (1976, pp. 34-35) propose to examine relevant literature, use expert panels and observe the way decisions are currently made in practice. Keeney (1992, pp. 55-69) proposes further techniques such as wish lists or taking different perspectives. Since the resulting list of objectives usually is very diverse, it is generally suggested (though not required by all MADA methods) to construct a hierarchy that organizes the different objectives by their scope, level of detail, etc. According to Brownlow and Watson (1987, pp. 310-311) this makes trade-off decisions more transparent, supports integration of experts for certain branches of the hierarchy and eases the evaluation process for the decision makers. Keeney (1992, pp. 69-86) discusses in detail how to structure objectives based on several real-life examples.

To construct a hierarchy, a top-down approach beginning with the broad overall objective or a bottom-up approach beginning with a collection of detailed objectives from the various stakeholders can be used (cf. Roy 1999, pp. 1-9). Any resulting hierarchy and the attributes used to evaluate an alternative's performance should possess a number of properties (cf. Keeney (1992), pp. 99-121; von Winterfeldt and Edwards 1986, pp. 36-45; Keeney and Raiffa 1976, pp. 41-53). Besides obvious requirements such as relevance, understandability and measurability of sub-objectives, a general requirement is that the hierarchy is mutually exclusive (no redundancies) and collectively exhaustive (completeness). Also, the decision maker should be able to express his preference for the various sub-objectives and complexity should be avoided whenever possible. As Brugha (1998) and Brownlow and Watson (1987) point out the problem structuring process can affect subsequent evaluation results and hence different approaches might have to be tested.

A critical assumption of all additive, compensatory MADA tools is that the decision maker's preferences with respect to one objective are independent of the alternative's attribute value with respect to another objective. This requirement is also referred to as *preferential independence* (cf. Belton and Stewart 2002, pp. 57-58) and should not be confused with statistical independence. Typical examples violating this assumption include the selection of a new car where the preference for a certain color is not independent of the type of car under consideration or the composition of a menu where a preference for a red or white wine is not independent of the type of main course (fish or meat) selected. While outranking methods do not by definition require preferentially independent objectives, Roy (1996, p. 229) points out the problems arising in their absence. Preferential independence can often be achieved by combining sub-objectives or setting minimum requirements for the options to be considered.

Of course, also the set of relevant alternatives has to be defined. If too many alternatives exist, a checklist approach might be used to arrive at a shortlist or a dominance analysis can be performed based on the performance matrix of the alternatives. In case of a lack of alternatives various techniques can be explored to generate additional alternatives (cf. Keeney 1992, pp. 198-240). The alternatives to be analyzed in detail should be described carefully because the way they are described can have unintended effects on decision makers' preference statements. The examples provided by Tversky and Kahneman (1986) illustrate this point.

Weight Elicitation Methods

Clearly, not all (sub-) objectives of a decision problem are equally important. Consequently, most MADA tools require that the decision maker defines weights for each objective. Common weight elicitation methods with varying degrees of sophistication are introduced below and pitfalls in weight elicitation discussed.

Simple Weight Elicitation Methods

Besides a direct estimation of weights other simple methods are commonly used in practice. The *weights from ranks* – approach (cf. Lackes 1983, pp. 386-387; Yoon and Hwang 1995, pp. 11-12) begins with an ordinal rank-ordering of all objectives and assignment of weight 100 (max 100 approach) to the most important or weight 10 (min 10 approach) to the least important objective. All other objectives are weighted relative to this reference point. With the *point allocation* method decision makers are asked to distribute a fixed number of points (often 100) across all objectives in a way reflecting their relative importance (cf. Schoemaker and Waid 1982, p. 184).

Simple Ratio Weighting (cf. Borcherding et al. 1991, p. 1608)

At each objective level the objectives are in a first step rank-ordered according to their importance by the decision maker. In a second step, a weight of 10 is assigned to the least important objective and all other objectives are judged as multiples of 10. Consistency is assessed by comparing the initial ordinal rank ordering with the ranks obtained from the ratio weighting.

Swing Weight Method (cf. Belton and Stewart 2002, pp. 135-139)

Starting with a theoretical alternative where all objectives have the worst possible attribute value, the decision maker is asked to change one attribute from worst to best beginning with the objective he considers most important. A value of 100 is assigned to this most important objective. The benefit of a swing from worst to best for all other objectives, rank-ordered by their importance, is assessed relative to this value.

Trade-off Weights (cf. Keeney and Raiffa 1976, pp. 121-123)

The decision maker compares two hypothetical alternatives that differ in two attribute values only. One alternative performs best on the first and worst on the second objective and the other vice versa. The decision maker chooses the preferred alternative thus deciding on which objective is more important. Subsequently, one alternative's performance is modified until the decision maker is indifferent between both alternatives. While only n-1 evaluations are required redundant comparisons allow consistency checks on the weights obtained.

Analytic Hierarchy Process Eigenvector Method (cf. Saaty 1980, pp. 165-197)

The pairwise comparison approach from the AHP method can also be used to assess objective weights in other MADA methods. As the method is discussed in detail in the context of the AHP, the reader is referred to Chapter 4.2.2 for further details.

Pitfalls in Weight Elicitation

Many empirical studies have been conducted to assess within-method consistency and between-method convergence of the various weight elicitation methods. Comparing simple weight elicitation methods, Bottomley and Doyle (2001) find that direct rating and max 100 show a higher within-method consistency than the min 10 approach. Borcherding et al. (1991) compare simple ratio weights, swing weights, tradeoff weights and pricing out weights.[40] Regarding within-method consistency they find that the simple ratio method fared best followed by the swing weight method and the tradeoff method. Weights obtained with the simple ratio and the swing weight method show a "reasonable" degree of convergence with the other methods performing worse. Comparing multiple regression, Analytic Hierarchy, tradeoffs, point allocation and unit weighting, Schoemaker and Waid (1982) find that with respect to within-method consistency on average all methods except unit weighting perform equally well while there are significant differences between the methods. Subjects in this study judged the AHP method to be at the same time the least complex and most trustworthy weighting method. Pöhönen and Hämäläinen (1998) compare several variants of the AHP method, simple (direct) weighting, swing weighting, simple ratio weighting and tradeoff weighting in an internet experiment. They find that 56% of participants are consistent between methods and 69% are consistent with respect to the highest ranked alternative. Other researchers did not identify significant differences between weights obtained using different elicitation methods (e.g., Eckenrode 1965). Hobbs (1980) uses the example of siting nuclear power plants to compare alternative weight elicitation methods and points out that the underlying tradeoff decisions as defined by the relative value of unit changes in the attribute value functions have to be considered. Hence, he recommends to use trade-off methods such as the indifference trade-off method.

Besides choice of method various behavioral effects can have an influence on the weights obtained (cf. Weber and Borcherding (1993) for a review). The most commonly cited issues include:

- The *attribute range effect* whereby attribute ranges are not sufficiently considered when determining weights (cf. von Nitzsch and Weber 1993).
- The *objective splitting effect* whereby the decomposition of an objective often leads to assignment of a higher overall weight compared to the aggregated objective (cf. Weber et al. 1988).
- The effects of a *hierarchical vs. nonhierarchical* objective structure whereby weights tend to be steeper in a hierarchical setting than in a non-hierarchical setting (cf. Stillwell et al. 1987).
- The effect of the decision makers' *reference point* whereby decision makers may assign different weights depending on the status quo perspec-

[40] Borcherding et al. also analyze external validity by comparing results obtained from tests with 200 students with weights provided by an expert panel but this approach seems to be questionable.

tive from which they analyze the problem (cf. Tversky and Kahneman 1991).

Obviously, science cannot provide clear guidance on which weight elicitation method to use but the AHP method appears to be a good choice. As weights critically influence the outcome of a decision analysis, decision makers should choose a method they feel comfortable with and keep in mind all assumptions, shortcomings and behavioral effects. Sensitivity analyses should be performed to assess the robustness of the solutions obtained and more than one method might be used to verify weights. If the latter approach is chosen one has to bear in mind that it is usually not possible to explain differences between the weights obtained from different methods.

4.2 Traditional MADA Methods

4.2.1 Simple Additive Weighting and Simple Scoring

Simple additive weighting (cf. Hwang and Yoon 1981, pp. 99-103) and simple scoring (cf. Lackes 1983; Zangemeister 1971) are probably the most widespread MADA methods in practice (cf. Stewart 1992, p. 573). Site selection applications have amongst others been reported by Hunt and Koulamas (1989) or Singhvi (1987, p. 52) and practically every textbook treating location decisions contains an exposition of simple scoring approaches (e.g., Jung 2004, pp. 71-73).

Overview

Application of simple additive weighting/simple scoring requires five-steps:

1. Identification of objectives and alternatives
2. Evaluation of alternatives
3. Determination of (sub-) objective weights
4. Additive aggregation of weighted partial preference values
5. Sensitivity analysis

The two methods only differ in the third step. Simple scoring uses a standardized interval scale (e.g., 0 – 100) for evaluation of all objectives. Decision makers directly rate the performance of each alternative with respect to each objective on this scale (cf. Fig. 25 for an example). Simple additive weighting uses direct rating on a standardized scale only in case of purely qualitative attributes. For numerical attributes scores are calculated by normalizing observed attribute values to match the standardized scale.

Two different types of scales can be used (cf. Belton and Stewart 2002, pp. 121-122). A *local scale* is defined with the best and worst existing alternatives forming the reference points for the bounds of the scale while a *global scale* is defined relative to absolute values for best and worst performance. Local scales are easier to define but global scales allow the decision maker to define objective weights independent of observed attribute values.

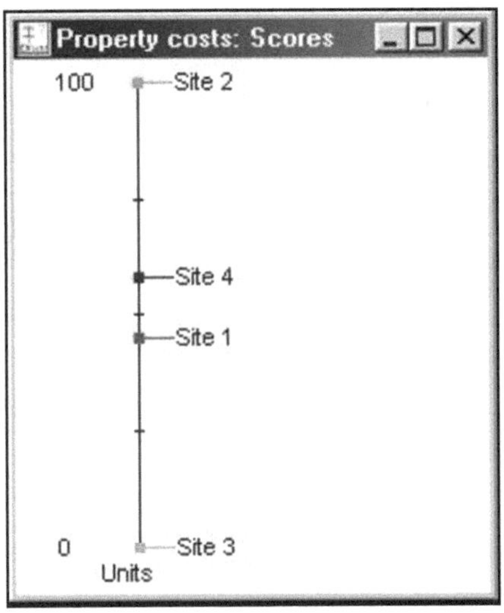

Fig. 25. Direct rating using VISA

The objective weights quantify the trade-off relationship between the sub-objectives. They can be obtained with any of the above-mentioned methods, but commonly a direct estimation is used with these simple methods. The combination of the simple scoring method with the ratio or swing weights approach is also referred to as Simple Multi-Attribute Rating Technique or SMART (cf. Goodwin and Wright 2004, pp. 27-58; von Winterfeldt and Edwards 1986, pp. 259-287).

Extensions to the Basic Method

Instead of directly scoring alternatives on a normalized scale or assuming a linear correlation between attribute values and decision makers' preferences an explicit value function can be constructed to convert observed attributes into preference values (cf. Goodwin and Wright 2004, pp. 37-39).

Keeney and Raiffa (1976, pp. 66-130) describe several ways of how to construct such value functions. If no directly measurable attributes exist for an objective, a scale that maps verbal descriptions with a numerical value indicating the decision maker's strength of preference has to be defined. The MACBETH (Measuring Attractiveness by a Categorical Based Evaluation Technique) procedure offers a pairwise comparison approach to support constructing such a scale (cf. Bana e Costa et al. 2005).

Using software tools such as VISA (Visual Interactive Sensitivity Analysis, cf. Belton and Vickers 1989) does not only make it more comfortable to analyze large-scale decision problems but especially allows the decision maker to conduct interactive visual sensitivity analyses. Bana e Costa et al. (1999) describe a case example that illustrates the interaction of the MACBETH approach with the VISA software.

Perceived Strengths

The key advantages of simple weighting/scoring are ease of use and understandability. Also, the evaluation itself does not require special software packages but can be performed with standard spreadsheet software. Special software tools such as VISA however contain features such as comprehensive sensitivity analyses, which are cumbersome to implement in spreadsheet software.

Criticism

Besides the general criticism of additive aggregation models the most criticized aspects about these simple methods are the fact that direct scoring of alternatives and direct estimation of weights often lead to results that lack an argumentative justification and that the resulting preference values lack an economic interpretation. Also, it is not possible to perform consistency checks on the decision maker's inputs.

4.2.2 Analytic Hierarchy Process

The Analytic Hierarchy Process, developed by Saaty (1980), is one of the most widespread MADA tools (see Vaidya and Kumar (2006), Forman and Gass (2001), Saaty and Vargas (2001), Vargas (1990), Partovi et al. (1990), Golden et al. (1989) or Zahedi (1986) for application reviews and Saaty (1986) for the axiomatic foundation). Case examples reporting the use of AHP in (plant) location decisions can for example be found in Min and Melachrinoudis (1999), Yang and Lee (1997), Kathawala and Gholamnezhad (1987), Yoon and Hwang (1985b) or Wu and Wu (1984).

Overview

The first step in using the AHP to analyze a decision problem is to hierarchically break down the decision problem (objective) into its constituent components and identify the alternatives to be evaluated. The resulting hierarchy consists of the overall objective (goal) and one or more levels of sub-objectives. The alternatives to be evaluated are added at the lowest level of the hierarchy. According to Saaty (1980, pp. 79-83) a cluster should not contain more than 7 elements because results from psychological tests show that 7+/-2 are the maximum number of elements a person can effectively compare simultaneously.

In a second step the alternatives' contribution to the various objectives and the relative importance of the objectives are assessed via pairwise comparison of each hierarchy cluster's elements. Usually, the 9-point ratio scale shown in Figure 26 is used to perform the comparisons. Since one assumption of the AHP is that the evaluations are reciprocal, n(n-1)/2 evaluations have to be completed to fill an n•n evaluation matrix.

Numerical rating	Verbal judgement
1	Equally important or preferred
3	Moderately more important or preferred
5	Strongly more important or preferred
7	Very strongly more important or preferred
9	Extremely more important or preferred
2,4,6,8	Intermediate values

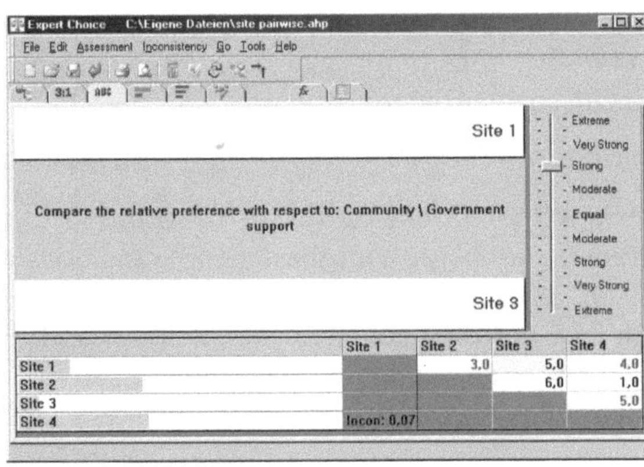

Fig. 26. AHP reference scale and example evaluation matrix (Expert Choice)[41]

Because the scale is bounded at value 9, elements to be compared should belong to a homogenous group (i.e., be comparable within one order of magnitude). If this is not the case, they have to be arranged in different clusters. Also, the decision maker should be aware of the consequences implied by use of a ratio scale. For example, assigning the value 5 (strong) in Figure 26 implies that site 1 is preferred five times as much as site 3 with respect to the sub-objective "government support". As attribute

[41] cf. Saaty (1980).

ranges affect the weight of the respective objective, Forman (1990a, p. 312) recommends that the evaluation be performed bottom-up so that attribute scales can be considered when eliciting objective weights.

In a third step the pairwise comparison matrices are synthesized to obtain weights/priorities. If the evaluations were required to be consistent, a simple row sum average would be sufficient (cf. Saaty 1990b, p. 19). However, a key characteristic of the AHP is to account for inconsistencies. The eigenvalue method used by the AHP estimates the weights of an inconsistent comparison matrix. The calculation process, for example described by Harker (1989), is iterative and thus requires a software tool. This software at the same time calculates a consistency ratio to check the degree of inconsistency in the comparisons (cf. Fig. 26). As a rule of thumb, Saaty (1980, p. 51) recommends that the comparisons should be revisited if the consistency ratio exceeds 10%. According to Belton (1986, p. 16) the objective weights in the AHP have to be interpreted as "the relative contribution of an average score (average over the options under consideration) on each objective".

Synthesis of the individual comparison matrices is the fourth step of the AHP. The evaluations obtained for the individual alternatives are multiplied with their respective relative importance and additively aggregated resulting in a linear, additive function to obtain overall preferences. In a group decision making environment, weights obtained from several decision makers are aggregated by calculating the geometric mean.

In a fifth step sensitivity analyses should be performed to explore how sensitive the obtained preference ranking of the alternatives is to changes in objective weights. This final step is especially relevant considering the limitations of weight elicitation discussed above. If the sensitivity analysis shows that the preference order changes already with small weight modifications the weights should be double-checked.

Extensions to the Basic Method

For cases where a large number of alternatives has to be evaluated, *absolute measurement* provides a second evaluation option that does not require a pairwise comparison of alternatives (cf. Saaty 2005, pp. 369-372). Also, methods to reduce the number of comparisons below n(n-1)/2 in large hierarchies while still ensuring a sufficient level of accuracy have been developed (cf. Millet and Harker 1990).

To allow for interdependencies and feedback between the various elements constituting a decision support model, Saaty (1996) extended the AHP into the more general *Analytic Network Process (ANP)*. A brief exposition of the ANP can be found in Saaty (2005, pp. 382-405) and Cheng et al. (2005) provide application examples. A case study of using the ANP for locating a repair-parts warehouse in electronics industry can be found in Sarkis and Sundarraj (2002).

Additionally, a scale that allows for comparisons exceeding one order of magnitude and functionality to include hard data are included in the Expert Choice software (cf. Forman and Selly 2002, p. 67 and pp. 140-144). Haines (1998) proposes approaches to accommodate cases where the decision maker is uncertain about his preferences and prefers to assign intervals of preference instead of single points.

Perceived Strengths

The huge number of application examples reported in literature demonstrate the versatility of the AHP. Goodwin and Wright (2004, pp. 420-421) provide further arguments for using the AHP. In their opinion the process of creating a hierarchy to model a decision problem contributes to a better understanding of the decision problem. The use of pairwise comparisons simplifies the judgmental task by focusing on two alternatives or objectives at a time. The redundancy resulting from requiring more than the minimum number of comparisons allows to check for inconsistencies while the eigenvalue method at the same time estimates weights taking into account remaining inconsistencies.

Another key advantage of the AHP is the availability of the sophisticated software package Expert Choice. In addition to the basic AHP functionality, the software supports a broad range of extra features (e.g., it allows for various forms of group decision making also via internet or voting devices). The use of Expert Choice to perform AHP evaluations is described in great detail in Forman and Selly (2002) which can be downloaded free of charge from the Expert Choice homepage.

Criticism

Several aspects of the AHP have been criticized in literature (see for example Goodwin and Wright (2004, pp. 421-422) for an overview). The most widely criticized aspect is the observed phenomenon of rank reversal (cf. Triantaphyllou 2001; Dyer 1990). If an irrelevant alternative is added or removed from the set of alternatives, a reversal of ranks can happen. To overcome this criticism, AHP has been extended and now offers two different synthesis modes: *ideal mode* and *distributive mode* (cf. Forman and Gass 2001, pp. 477-481; Saaty and Vargas 2001, pp. 42-43; Saaty 1994, pp. 439-444). The distributive mode assumes that judgments are distributed amongst alternatives (similar to an allocation of scarce resources). If an alternative is added or the score of an alternative is changed, the priorities of the other alternatives change, too. The rank reversal that may occur in certain cases is considered to be consistent with the underlying assumption of scarcity. Contrarily, the ideal mode calculates preferences/priorities relative to the best score as a fixed benchmark by dividing the priorities of the alternatives by the largest priority among them. This way, a new alternative is compared against the ideal alternative for each objective and no rank reversal can occur. Alternative methods to prevent rank reversal can for example be found in Schoner et al. (1993).

Also, the eigenvalue method has been criticized because of its perceived lack of transparency. Amongst others, Belton and Stewart (2002, pp. 158-159) have suggested to use logarithmic least squares (that do not require an iterative calculation process) instead. Saaty (1990a) however strongly defends the eigenvalue method because of the way it deals with inconsistent judgments.

Criticism of the 9-point scale used in the AHP includes aspects such as its being bounded, lacking theoretical validation for the verbal equivalents used and especially the ratio scale property (cf. Belton 1986, p. 11). In practice, it can be observed that the values between 7 and 9 are only infrequently assigned because of a perceived lack of difference to the values 5 and 6. Alternative ratio scales and their performance relative to the traditional AHP scale can for example be found in Leskinen (2000) and Pöhönen and Hämäläinen (1998) while Schoner and Wedley (1989) suggest to use the *magnitude estimation* method instead.

4.3 Outranking Approaches

Outranking methods were developed to provide MADA methods that are not based on the strong assumptions underlying the classical approaches while at the same time providing insights exceeding a simple dominance analysis (cf. Vincke 1992, pp. 57-58). Consequently, no additive, compensatory aggregation procedure is used and alternatives may be classified as being incomparable. Below, the most common variants from the PROMETHEE and ELECTRE families of outranking methods are introduced. Vincke (1999) presents further methods and additional references.

4.3.1 ELECTRE

The ELECTRE (ELimination Et Choix Traduisant la REalité) family is the first major family of outranking methods. Individual methods within the family differ with respect to their complexity, the richness of information required and the type of decision situation they can be applied to. ELECTRE I – III are introduced below to illustrate the concept. An introduction to the whole family of methods can be found in Figueira et al. (2005b) and Vincke (1992).

Overview

Applying ELECTRE methods comprises two major steps: constructing outranking relationships between the alternatives via pairwise comparison and evaluating these relationships to arrive at a recommendation (cf. Figueira et al. 2005b, p. 138). From the pairwise comparisons the decision maker obtains a concordance index supporting the hypothesis that an alternative *a* is at least as good as *b* and a discordance index measuring the strength of the evidence against this hypothesis. The evaluation is adjusted for the relative importance of the sub-objectives via weights (the "voting power" of each sub-objective) and veto thresholds that indicate from which level onwards the discordance index vetoes an outranking relationship (cf. Roy 1990b, pp. 167-168).

ELECTRE I, while being of limited applicability in practice, was the first method developed and illustrates the approach (cf. Figueira et al. 2005b, p. 140). It calculates the concordance index of an alternative *a* over an alternative *b* by dividing the sum of all sub-objectives' weights where *a's* performance is at least as good as *b's* by the sum of all weights. The discordance index was originally defined as the maximum difference between the performance levels of two alternatives for all sub-objectives where *b* outperforms *a*. If this difference exceeds the veto threshold, *a* cannot outrank *b*. As this definition of the discordance index requires a cardinal scale and comparability of the scales across all sub-objectives, less demanding approaches that use individual veto thresholds for each sub-objective were proposed (cf. Belton and Stewart 2002, p. 236).

To build the outranking relationship, the decision maker has to define concordance and discordance thresholds. An alternative *a* outranks *b* if the concordance index is greater than or equal to the concordance threshold and the discordance index is less than the discordance threshold. The decision maker may have to vary these thresholds in order to obtain an informative outranking relation. The result is a set of preferred alternatives such that any alternative not in the set is outranked by at least one alternative in the set and all alternatives in the set are incomparable.

Extensions to the Basic Method

The ELECTRE family has been extended frequently to improve the original method and develop methods for different applications. ELECTRE II aims at ranking alternatives. It constructs a downward and an upward ranking in a multi-step procedure based on a strong and a weak outranking relationship and consequently requires two pairs of concordance and discordance thresholds. To better model the decision maker's preferences in the pairwise comparison process, ELECTRE III introduces quasi-criteria based on indifference and preference thresholds. For a detailed discussion of these two methods refer to Belton and Stewart (2002, pp. 241-250).

Further methods include ELECTRE TRI, which supports a classification of alternatives into various categories and ELECTRE IV, which can be used in settings where the decision maker cannot specify sub-objective weights (cf. Figueira et al. 2005b, pp. 146-149).

Perceived Strengths

Hwang and Yoon (1981, p. 127) consider the ELECTRE family of methods to be "one of the best methods" because of the way it utilizes the information available in the decision matrix and a logic they consider to be simple.

Criticism

Belton and Stewart (2002, p. 239) criticize that with the ELECTRE methods it is not possible to perform sensitivity or robustness analyses on the weights and the thresholds in an automated or interactive way. Also, they point out that weights and thresholds have no clear interpretation in terms of the decision maker's preferences and thus are often modified in an ad-hoc manner to obtain adequate outranking relations (Roy (1990b, pp. 176-179) discusses how to assign numerical values to these parameters). Additionally, Belton and Stewart (2002, pp. 246-247) criticize that the aggregation procedure of ELECTRE II and III is difficult to communicate and that with ELECTRE III counterintuitive results and rank reversal may occur if an alternative is added or removed.

In practice, the applicability of ELECTRE methods is also limited by the lack of commercial software packages. The software tools available are academic developments by the team around Bernard Roy from LAMSADE (cf. Weistroffer et al. 2005)

4.3.2 PROMETHEE

The PROMETHEE method (Preference Ranking Organization METHod for Enrichment Evaluations) method was originally developed by Brans et al. (1984). Pavic and Babic (1991) provide an example of applying PROMETHEE for plant selection and Karkazis (1989) uses the method for locating facilities in a competitive environment.

Overview

PROMETHEE evaluates discrete alternatives based on pairwise comparisons for each sub-objective. However, these comparisons are not performed "manually". Instead, partial preference functions, usually using the difference between two alternatives' attribute values as data input, are used to perform the pairwise comparisons. These partial preference functions have to be defined for each sub-objective as a first step of using PROMETHEE. They are used to determine the decision maker's strength of preference for an alternative (with 0 assigned for indifference, 1 for strict preference and values between 0 and 1 for intermittent preference values).

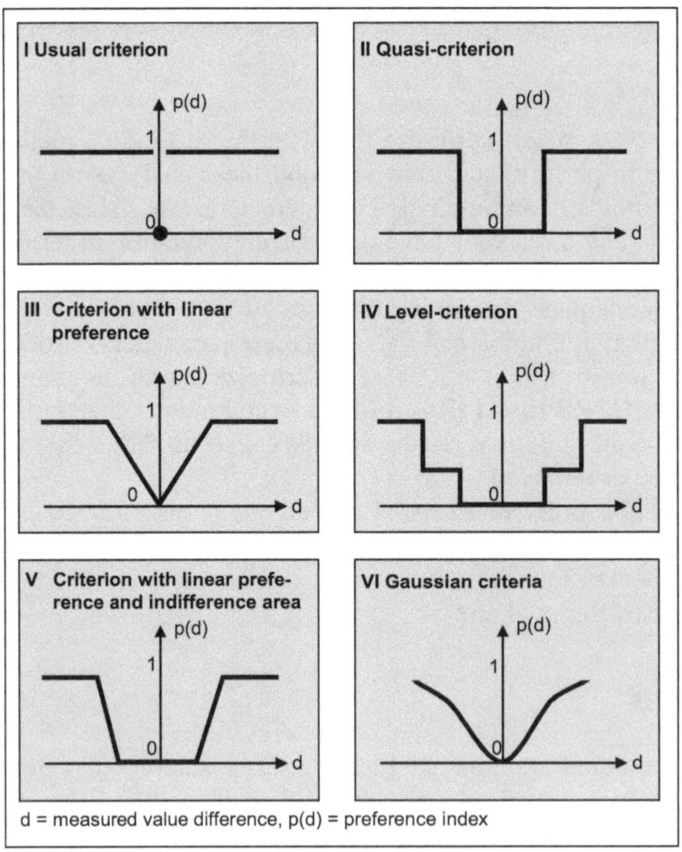

Fig. 27. Generalized PROMETHEE criteria[42]

[42] Source: Brans et al. (1986), p. 231.

It is recommended to use one of the six generalized preference functions shown in Figure 27 (cf. Brans and Vincke 1985, pp. 649-652). Depending on the selected function, the decision maker has to define indifference, preference and Gaussian thresholds to specify the respective function. For purely qualitative criteria the decision maker has to define a rating scale (e.g., 3-point or 5-point scale) that transforms the decision maker's assessment of each alternative into a numerical value used for the pairwise comparisons. As a result of the comparisons, the decision maker obtains a comparison matrix for each sub-objective providing a positive (row sum) and negative (column sum) outranking value for each alternative.

In a second step, these partial outranking values are aggregated into each alternative's total positive and negative outranking values by calculating the weighted sum of the partial values. As no specific method is recommended to determine criteria weights, any of the methods introduced above can be used. When defining weights the decision maker should keep in mind their meaning. For the non-compensatory parts of the method, the weights have to be interpreted as number of votes while for the compensatory parts they define tradeoffs between degrees of outranking or preference (cf. Vincke 1999, pp. 11-18). However, the weights should be obtained only after having evaluated the alternatives so that the decision maker can consider attribute ranges when defining weights.

Two different approaches towards synthesizing the results into a ranking of the alternatives considered can be applied in the third step. PROMETHEE I is a partial ranking based on an analysis of dominance relations between alternatives. Using the weights defined for each criterion, the positive outranking value P^+ shows how much an alternative dominates all other alternatives (concordance principle). The negative outranking value P^- shows how much it is dominated by other alternatives (discordance principle). If $P^+(a) > P^+(b)$ and $P^-(a) \leq P^-(b)$ or $P^+(a) \geq P^+(b)$ and $P^-(a) < P^-(b)$ then alternative a dominates alternative b. By comparing all alternatives in this fashion, the decision maker can obtain a partial ranking. In this partial ranking it is also possible that alternatives are found to be incomparable (e.g., alternative a has a stronger positive outranking value than alternative b but also a higher negative outranking value, cf. Brans et al. 1986, p. 233). If a complete ranking is required, the decision maker can resort to PROMETHEE II. This method uses the net preference index $P^+ - P^-$ to rank all alternatives. Figure 28 shows a partial ranking with incomparability and a complete ranking obtained using the Decision Lab software.

In a final step, again sensitivity analyses should be performed with respect to the influence of the weights on the ranking obtained. The software tool Decision Lab 2000 that implements PROMETHEE allows to determine stability intervals that show the range within which a weight can be varied without changing the final ranking.

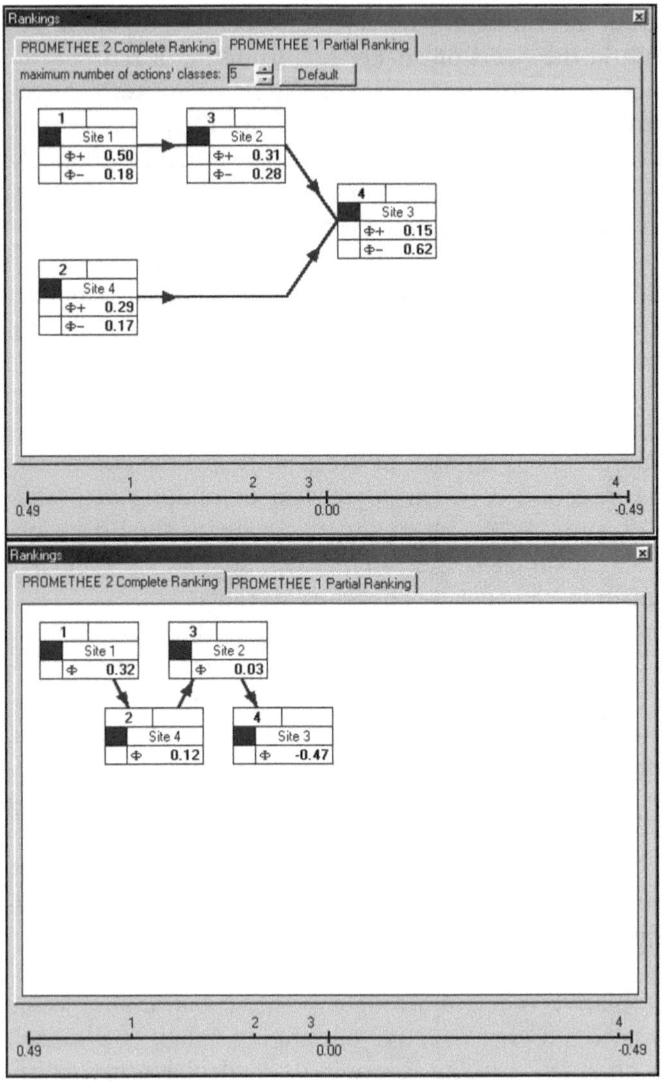

Fig. 28. Partial and complete rankings from Decision Lab 2000

Extensions to the Basic Method

Various extensions to the original method have been developed. These include interval-based and continuous rankings, a method to identify a subset of alternatives satisfying certain constraints and a group decision support procedure. The GAIA plane allows a further visual, interactive analysis of a decision problem by projecting a maximum amount of information into a meta-model. An overview of the various extensions can be found in Brans and Mareschal (2005). Additionally, Goumas and Lygerou (2000) have developed a variant of PROMETHEE for fuzzy decision environments and a specific application for financial institutions was implemented in the "Bankadvisor" software (cf. Brans and Mareschal 1990, pp. 242-250).

Perceived Strengths

Besides the core features of all outranking methods, use of PROMETHEE offers a range of advantages. The generalized preference functions enable the decision maker to flexibly model his preferences for each sub-objective while the various extensions ensure the applicability in a broad range of decision situations. The GAIA plane (Graphical Analysis for Interactive Assistance) allows an interactive exploration of the solution space to better understand the structure of the decision problem. With the Decision Lab software a powerful tool is available that supports not only the original PROMETHEE methods but also the various extensions (see Geldermann and Zhang (2001) for a review). It also allows the decision maker to perform various sensitivity analyses and provides a broad range of reports and graphics to communicate the results of an analysis.

Criticism

Defining partial preference functions which require up to three different threshold parameters for each sub-objective makes using PROMETHEE a relatively complex task. Estimating these parameters might be difficult/impossible in many decision environments. Even if the parameters can be specified it is often hard to provide a sound analytical foundation for doing so. Additionally, partial preference rankings are more difficult to understand than additive preference functions and hence the results obtained are more difficult to communicate to stakeholders.

4.4 Data Envelopment Analysis

Data Envelopment Analysis (DEA) is used to comparatively evaluate "decision making units" (DMUs) that convert inputs into outputs and identify sources and levels of relative inefficiency for every DMU with respect to each input and output (cf. Cooper 2001, p. 183). The approach was originally developed by Charnes et al. (1978). DEA has received considerable attention in literature. The bibliography in Seiford (1994) already contains more than 450 publications. While it is used primarily for efficiency analyses, the use of DEA as a multiple criteria decision analysis tool has been propagated by several authors (cf. Sarkis 2000; Doyle and Green 1993) and applications to site selection problems have been reported (e.g., Thompson et al. 1986).

Overview

DEA builds on the classical definition of efficiency: maximize output with given inputs or minimize inputs required for a given output. Instead of using a single, aggregate measure of input and output to rank the efficiency of the DMUs, a set of inputs and outputs is considered simultaneously without requiring these factors to be measured in a single dimension such as monetary units. The objective of DEA is to determine for each DMU whether or not it operates efficiently relative to its peers, given its inputs and outputs. A DMU is found inefficient if another DMU or a convex combination of other DMUs is able to produce the same or better results for the various outputs with fewer required inputs.

Using DEA requires three major steps. In a first step, the DMUs to be considered in the analysis have to be determined. These have to be homogeneous with respect to their tasks and the types of inputs and outputs used. The second step is to select the input and output factors that are to be used for the comparison. DEA also accommodates qualitative factors with the only requirement being that numerical values have to be assigned to all factors. Finally, a mathematical programming model such as the basic formulation provided below is solved separately for every DMU.

Table 21. DEA indices, parameters and decision variables

Indices			
$i \in I$	Input factors	$o \in O$	Output factors
$u \in U$	Decision Making Units (DMUs)	e	DMU being evaluated
Parameters			
a_{iu}	Observed use of input factor i at DMU u		
A_{ou}	Observed output of output factor o at DMU u		
δ	Small positive number $(0 < \delta << 1)$		
Decision variables			
s_i^-, s_o^+	Slack variables for input and output factors		
π_u	Weights of virtual DMU		
E	Measure of efficiency for DMU e		

$$Min\ E - \delta \cdot \sum_{i \in I} s_i^- - \delta \cdot \sum_{o \in O} s_o^+ \tag{4.1}$$

$$\sum_{u \in U} A_{ou} \cdot \pi_u - s_o^+ = A_{oe} \qquad \forall o \in O \quad (4.2)$$

$$\sum_{u \in U} a_{iu} \cdot \pi_u + s_i^- = E \cdot a_{ie} \qquad \forall i \in I \quad (4.3)$$

$$E, s_i^-, s_o^+, \pi_u \geq 0 \qquad \forall i \in I, o \in O, u \in U \quad (4.4)$$

The objective function (4.1) minimizes the efficiency measure E. For the smallest E obtained the slack variables are maximized. This objective hierarchy is achieved by including the "very small" parameter δ that subordinates the maximization of the slack variables under the minimization of E. Equations (4.2) and (4.3) specify the output and input factor comparisons. The slack variables contain the surplus of output factors and underconsumption of input factors respectively as compared to the virtual DMU. The weight parameters π_u are determined by the optimization model and describe the linear combination of real DMUs constituting the virtual DMU. Restriction (4.4) contains non-negativity constraints.

The DEA model estimates an empirical production function which achieves the highest value of outputs that could be generated based on the input-output vectors of the DMUs analyzed. The efficiency of an individual DMU is measured by the distance of its input-output combination to this production function. An individual DMU is *enveloped from above* if the model can identify a combination of other output vectors for the same input vector that is at least as good as the one of the DMU considered for all output factors. Analogous it is *enveloped from below* if the model can identify a combination of other input vectors for the same output vector that requires less than or the same amounts of inputs as the one of the DMU considered. If a DMU cannot be enveloped by a combination of other DMUs it is efficient. In this case the measure of efficiency *E* takes on a value of 1 and the slack variables are zero. For inefficient DMUs the value of the efficiency measure indicates the extent to which all outputs could be increased or all inputs be decreased and the slack factors provide the absolute units by which specific inputs could be decreased/outputs increased in addition to the general increase/decrease if the DMU were to be brought to efficient performance levels. These improvements are only indicative of potential improvements because the projection to the efficient frontier can also be based on a virtual DMU.[43]

[43] This paragraph is based on Charnes et al. (1994, pp. 6-7) and Golany and Roll (1989, p. 248). Golany and Roll also provide a detailed discussion of how to apply DEA and numerous examples.

Extensions to the Basic Method

Numerous alternative formulations of the optimization model underlying DEA have been developed for example to take into account the effects of *returns to scale* or to link DEA to *Cobb-Douglas* type production functions (cf. Golany and Roll (1989, p. 249), Lewin and Seiford (1997, p. 3) or Charnes et al. (1994, pp. 23-47) for additional models). Furthermore, constraints can be added to the basic models for example to incorporate known relationships between the weights of inputs, an assumed directional relationship between outputs or to impose bounds on the weight range available for certain factors (cf. Charnes et al. 1994, pp. 49-61; Golany and Roll 1989, p. 245). Since the basic DEA models only discriminate between efficient and inefficient DMUs, different techniques to derive complete rankings of the DMUs have been proposed. While some of these only provide a ranking within the set of efficient DMUs (e.g., the super-efficiency approach originally proposed by Andersen and Petersen (1993)) others such as statistics-based models provide a ranking of all DMUs considered (e.g., Sinuany-Stern and Friedman 1998). Adler et al. (2002) provide a review of various approaches. Since the different techniques draw on different assumptions and lead to different rankings of the DMUs they conclude that a selection based on the application case is required.

Perceived Strengths

According to Golany and Roll (1989, pp. 246-247) the main strength of DEA is that it can be used to assess the efficiency of DMUs in cases where no functional relationship between inputs and outputs can be defined up front and the weights of the various factors are not well-defined. Another strength pointed out by Bouyssou (1999, p. 974) is that no subjective or economic parameters are required to perform the analysis.

Criticism

A major criticism of DEA is that due to the relatively complex underlying mathematical calculations the extent of analytical skills required to understand the approach often exceeds that found with decision makers (cf. Sarkis 2000, p. 555; Belton and Vickers 1993, p. 885). Furthermore, DEA only assesses relative efficiencies within the group of DMUs considered and results depend on the factors included in the analysis and the numerical values assigned to factors not directly measurable (cf. Golany and Roll 1989, p. 247). If weights are not bounded, a result indicating that a DMU is efficient can be associated with an extreme distribution of weights. The extension to use bounded weights addresses this problem but Belton and Vickers (1993, p. 886) criticize that as the meaning of these bounds is not clear it is hard to set them and interpreting the results obtained becomes even more difficult.

Use of DEA as a Multiple Criteria Decision Analysis Tool

As Stewart (1996, p. 663) points out, DEA can be used for selection/ranking problems usually considered to be MCDA problems if the (sub-) objectives can be divided into costs/inputs and benefits/outputs. Doyle and Green (1993) argue that DEA is well suited for such MCDA problems especially because it has the advantage of not requiring weights and preference judgments (while being able to integrate them if desired) and the fact that the data inputs are easy to understand. DEA can also be integrated with a MCDA method for example to screen alternatives prior to weight and preference elicitation (cf. Stewart 1996, p. 664) or by using the AHP to derive complete rankings based on the data obtained from an initial DEA analysis (cf. Sinuany-Stern et al. 2000). Sarkis (2000) compared the results obtained using DEA with the results from several MADA methods such as SMART, PROMETHEE and ELECTRE. He concludes that despite some disparate ranking results DEA performs well as a MADA tool and that incorporating decision makers' preferences in the DEA analysis leads to greater correlation of the results with those obtained from value-based methods.

Taking an opposite position, Bouyssou (1999, pp. 975-976) is critical of using DEA for selection and ranking problems because DEA does not discriminate between efficiency and convex efficiency leading to discrimination problems in the selection case and arbitrary rankings of convexly dominated alternatives. Furthermore, the ranking DEA returns only consists of each DMUs' efficiency measure E. A direct interpretation of the values is impossible because it is related to the virtual DMU which is different for every DMU. Consequently, trade-off analyses to qualitatively assess the different performance levels via drill-downs into a subset of the criteria (inputs) are not supported.

4.5 A Specialty Chemicals Site Assessment Model

In the course of this work a site assessment model was developed together with the industrial partner cooperating in the research effort. Attained insights are provided in this chapter in a generalized form to preserve confidentiality.

4.5.1 Choice of Method

Selecting the most appropriate decision support methodology is no trivial task. Ozernoy (1992) even developed an expert system to support selecting the "best" method. Following Stewart (1992, p. 569) an assessment of the various MADA methods introduced above should be based on ease of use by non-experts, transparency of the method to the decision maker and clarity of the meaning of the inputs required. Ozernoy (1992, p. 160) adds that the decision maker needs to be capable of providing the preference information required by the method. Additionally, for application to large-scale site selection and ranking problems adequate software support (for an overview of available software cf. Weistroffer et al. 2005) and the ability to perform both selection and clustering/ranking tasks is required.

The first step is to determine whether a classical additive model, an outranking approach or DEA should be used. The criteria to be used for the site selection/site ranking task do not lend themselves to a clear classification into input and output parameters. Furthermore, the relatively complex mathematical procedure underlying DEA was perceived to be inappropriate for support of top management decisions that are to a large extent driven by qualitative factors. Hence, DEA was ruled out first.

The choice between the classical additive models and the outranking approaches is predominantly a tradeoff between the sometimes limiting assumptions underlying a fully compensatory additive model and the many parameters required for specifying outranking methods. Especially with the ELECTRE methods, the various thresholds required are difficult to explain let alone elicit. With PROMETHEE the parameters needed to specify the value functions underlying the method are more intuitive. However, qualitative criteria have to be converted into numerical values, e.g., via a rating scale, the aggregation method is relatively complex and a separate method is required to derive criteria weights. Therefore, in practice the decision analyst often has to determine these parameters (cf. Sarkis 2000, p. 544) which could increase the reluctance of the decision makers to accept the results obtained. Due to the more complex nature of outranking methods, Belton and Stewart (2002, pp. 258-259) also do not recommend their use in workshop settings where the decision analyst works directly with the decision makers. Yet, the ability to use the support tool in a live setting together with decision makers was seen as a key requirement (see Phillips (1986, pp. 193-194) for an example of how to set up group decision workshops). The above-mentioned practical disadvantages of the outranking approaches led the decision makers from industry the author discussed the alternative options with to prefer an additive model. Sensitivity analyses on the weights were perceived to be sufficient to ensure that the compensatory nature does not lead to undesired results.

Thus, the choice was narrowed down to one of the scoring/value function methods versus the AHP. Stewart (1992, p. 575) and Belton and Stewart (2002, p. 159) tend to favor the scoring/value function methods because they are considered to be easier to understand than the AHP ratio scale and eigenvalue procedures. Belton (1986, p. 18) does not give a recommendation but provides anecdotal evidence indicating that decision makers consider the former to be more transparent and easier to communicate. Contrarily, Schoemaker and Waid (1982) found that test subjects consider the AHP method to be both easy to use and trustworthy.

The author's experience has been that decision makers from industry find the pairwise comparison process underlying the AHP and the consistency checks it contains intuitively appealing and a good basis for discussing the advantages of the various alternatives. The details of the eigenvalue method have not been of interest in practice while the group decision functionalities, the direct rating mode and the sensitivity analyses provided by the Expert Choice software have been strong arguments in favor of the AHP. In the course of the project, the AHP method has been applied successfully to the ranking of all existing sites of a multinational specialty chemicals company, the selection of an individual site from a set of alternatives and the evaluation of different restructuring options within a country. The decision makers involved have been satisfied both with the process of analyzing the decision problems based on the AHP and the insights gained from the resulting evaluation. Consequently, while each decision maker has to decide individually which method he prefers, use of the AHP can clearly be recommended in the context of site selection/ranking.

4.5.2 The AHP Site Assessment Model

The objective was to develop a site assessment model based on the Analytic Hierarchy Process that can be used both for site selection and site ranking/controlling purposes. While the different application situations to some extent require different criteria the intention was to develop a model as uniform as possible to achieve consistency in site strategy processes.

Creation of the Objectives Hierarchy

The term location factor was already introduced in Chapter 2.3.2 and a few examples were provided. To construct the objectives hierarchy for assessing chemical production sites, findings from empirical research containing data on location factors specific to chemical industry are a good starting point. Bass et al. (1977) analyzed the importance of 44 location factors in 118 international plant location decisions of which 21 belong to chemical industry. Roughly summarizing their findings it can be said that factors such as access to markets and suppliers, transportation infrastructure, site development costs, environmental regulations and host nation trade policies were considered very important by chemical companies while factors such as media and advertising services, quality of living, public services, labor market and host government characteristics were deemed to be much less important. In a comprehensive empirical study that included 108 participants from chemical industry Brede (1971) similarly found that labor availability, cost and size/quality of industrial properties and room for future expansions are important location factors while local market potential and transportation costs have only limited and proximity to competitors a negligible influence on location decisions. In a case-study Haigh (1990, p. 27) found that seaport access and distance from residential areas can be further important factors in chemical industry.

Since more comprehensive and/or recent empirical data on the importance of different location factors for chemical industry was not available, a list containing a broad range of potential location factors was compiled in workshops with experts from all businesses of the company using the above-cited location factors as a starting point. The location factors were clustered in a hierarchy and an initial prioritization was performed using the Expert Choice software. In a subsequent workshop the priority scores obtained were discussed and the objectives hierarchy was streamlined to reduce the complexity of the resulting model. Figure 29 contains an exemplary objectives hierarchy containing pertinent location factors.

Compared to the empirical findings reported above, the relative importance of trade policies and market access has decreased significantly due to the trade liberalizations that were implemented in the meantime. Instead, today factor cost differences and access to raw materials and reliable utilities drives many location decisions in specialty chemicals. With respect to environmental regulations today most multinational companies implement uniform standards across their global networks which typically exceed the requirements of countries with laxer regulations. The perceived relative importance the industrial partner assigned to the location factors, which depends on the application context and cannot be generalized, is reported in Chapter 4.5.3.

Depending on the application context only a subset of the location factors listed in Figure 29 is used. For example, in a value chain production network design project business strategy considerations and quantitative location factors are considered in the network design phase. Hence, typically only a qualitative site assessment is required in the site selection phase and the subset of factors shaded in light gray is used. The location factors shaded in dark grey are only required if existing sites (in the form of joint ventures, purchase of a site or co-location at a site already operated by the company) are considered. Also, further location factors might have to be included to reflect specific requirements of the type of value chain considered. For site ranking purposes the strategic assessment of the businesses present at a site derived from corporate strategy considerations is included to integrate site strategy with business strategy. An aggregation into a meaningful site ranking is discussed below.

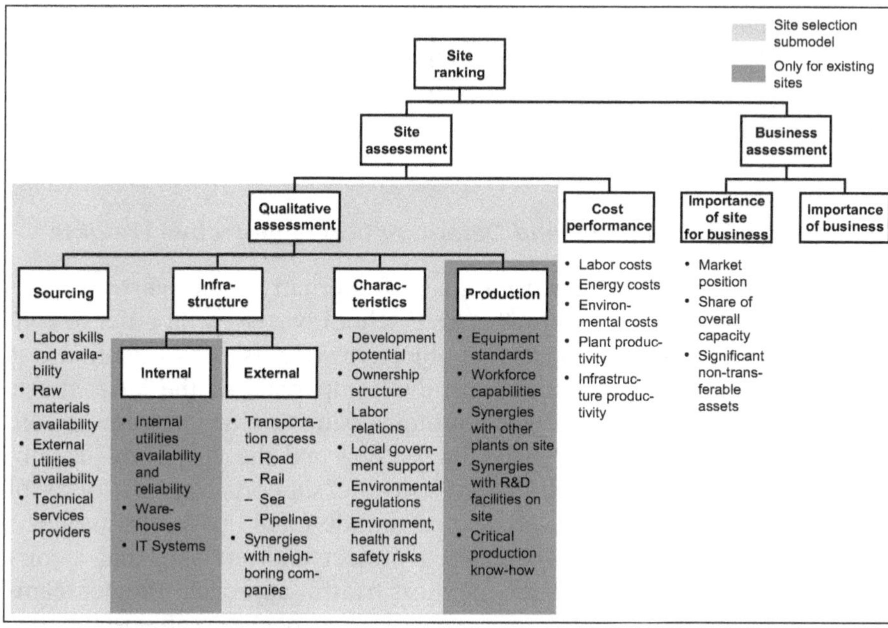

Fig. 29. Objectives hierarchy for site assessment

The majority of the location factors shown in Figure 29 are self-explanatory. While no uniform scale is required when using the pairwise comparison approach in a site selection situation, it has proven helpful to create a joint definition of each attribute that also contains an indication of minimum requirements and ideal state together with all stakeholders involved in the site selection process. To this end, verbal scale descriptions were defined for all sub-objectives. The global scale to use the model for site controlling (see below) was then created based on these definitions.

The sourcing objectives only concern the availability in sufficient quality and quantity of major input factors because prices are either considered in the network design model or as part of the cost performance branch. Ownership structures affect the degree of control the company can exert. For this reason, minority joint ventures and locating at industrial parks are often less preferred alternatives. Labor relations typically concern the influence of unions but the degree of work-time flexibility and the intrinsic motivation of employees are also important elements to consider. In the cost performance branch two productivity Key Performance Indicators (KPIs) are included that aggregate the cost performance of a site's production and infrastructure units. These KPIs have to be determined via company-internal benchmarking or use of DEA to be comparable across sites. As country selection takes place in the network design phase, country risks, which are often included in qualitative site assessments, already have to be considered when selecting potential investment locations for the network design model. Appendix 3 contains an objectives hierarchy to assess country risk that could be used to directly integrate country risks into the model.

Evaluation of Alternatives and Determination of Objective Weights

For selection of a single site from a set of alternatives the pairwise comparison approach propagated by the AHP school was used. In order to consider the effect of attribute ranges on objective weights, the evaluation was performed bottom up beginning with the comparison of the alternatives. Ideally, the project team conducts a group session to reach agreement both on the relative performance of the alternatives and the weight of the subobjectives. While the Expert Choice software supports various forms of group decision making, in the context of an individual investment project a setting where a facilitator moderates a project team session and a joint evaluation is performed has been the most fruitful approach. Project teams should be staffed with representatives from the business affected by the decision, a member of the corporate (site-)strategy team and members of the country management of the target country. To ensure the assessment of the alternatives is consistent with the overall site strategy, the global scale defined for site rankings (see below) can be used to provide guidance with the comparisons.

For site rankings the large number of sites to be included requires a ratings approach instead of the pairwise comparison of alternatives. To this end a global scale that can be used independently of the alternatives considered was defined (cf. Fig. 30). As more information is available for existing sites, it is desirable to use KPIs also for qualitative factors[44] whenever possible. For example, based on historical data the average percentage of raw material deliveries supplied "on time in full (OTIF)" can be used to assess raw material availability. To convert available KPIs into preference data value functions were defined (cf. Fig. 31). With other sub-objectives the rating was driven by corporate policies. For example, the cooperation partner preferred not to locate plants at industrial parks despite the smaller share of fixed costs associated with this option because of the higher degree of control they can exert with full ownership of a site. Finally, some criteria such as degree of government support had to be rated subjectively by local management and in cases of missing or imprecise data expert judgments were used.

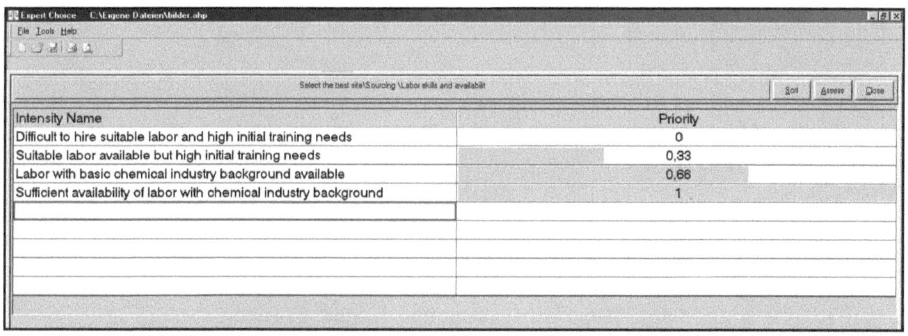

Fig. 30. Ratings approach in Expert Choice

The AHP model was directly linked to a site information system database where a corporate team compiled the required data. Whenever no value functions were used, the priorities associated with rating categories were either entered directly (e.g., in 0.25 intervals for a four-point scale) or assessed via pairwise comparisons of the scale descriptions. Objective weights were determined using the regular pairwise comparison approach considering the attribute ranges implied by the global scale previously defined. To account for the fact that objective weights can differ between the business segments of a company representatives from all businesses individually assessed weights for the site ranking model.

[44] Which were defined as having no directly measurable financial impact, cf. p. 23.

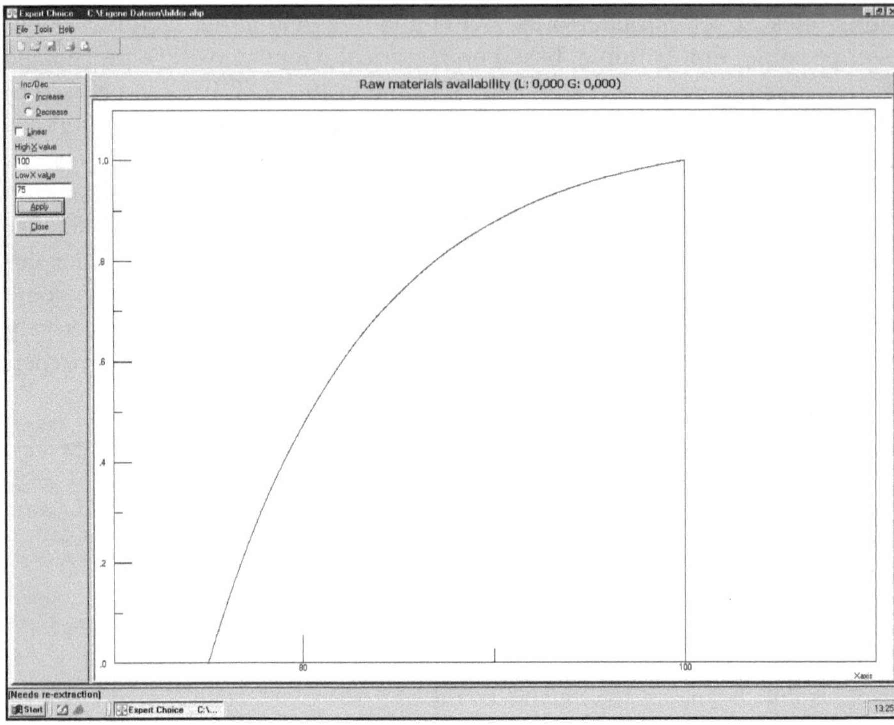

Fig. 31. Rating and value function approach in Expert Choice

Synthesis of Results and Sensitivity Analyses

In site selection applications once the evaluation is completed, a synthesis is obtained from the model indicating the preferred alternative. Especially in cases where several alternatives show similar preference levels, a detailed dissection of the results is required. To this end, both the performance profile of the sites and the sensitivity of the evaluation results to weight changes for major location factors should be analyzed. The Expert Choice software offers a broad range of features to facilitate this task. Figure 32 contains four illustrative examples. The upper left corner shows the performance profile of all alternatives for one level of the objectives hierarchy and the upper right corner the sensitivity of the overall results to weight changes for a single sub-objective. The feature shown in the lower left corner allows to interactively explore the effects of weight changes on the evaluation results and the feature in the lower left corner provides a direct comparison of two alternatives.

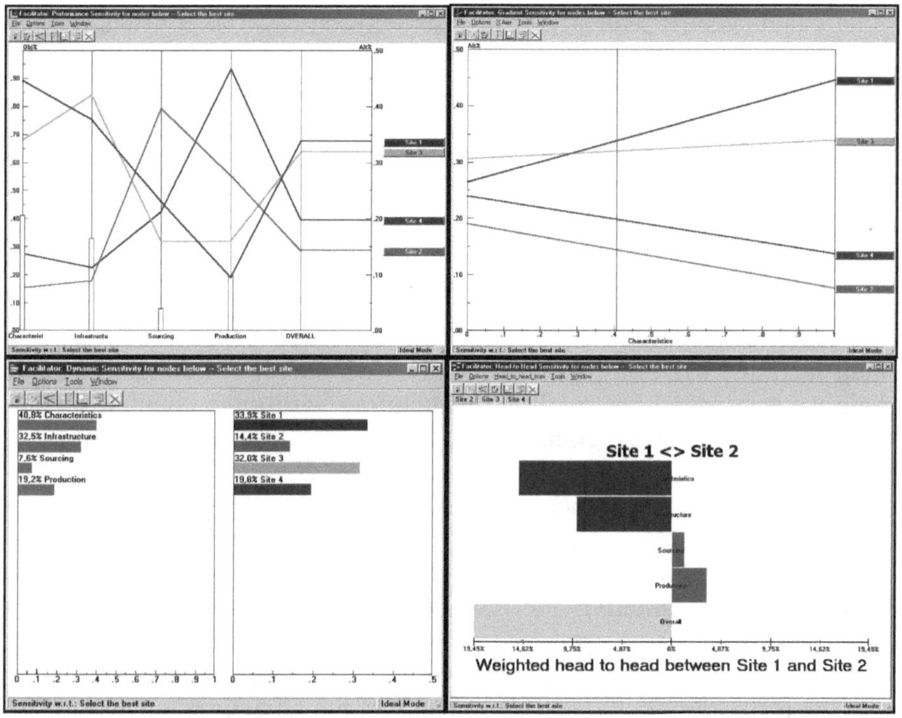

Fig. 32. Expert Choice sensitivity analysis examples

Additionally, if cost differences between alternatives exist that were not considered in the network design phase, potential trade-offs between quantitative and qualitative performance of the alternatives should be analyzed.

For site rankings a different synthesis approach is required to account for those sites that host plants from more than one value chain. For the site assessment branch of the hierarchy, business-specific weights were determined instead of uniform weights. The group decision analysis features of Expert Choice easily allow for a clustering of the evaluators into separate groups and an aggregation of the weights for a subset of the participants. Consequently, the production and business managers who participated in the assessment were clustered based on the value chains they represent. Similarly, the business assessment branch is evaluated separately for each value chain present at a site. It combines an assessment of the site's role for the value chain (e.g., center of excellence, share of overall production capacity) with the overall strategic importance of the value chain as defined by corporate strategy. This way a value chain-specific ranking is obtained for every site. Aggregate site rankings are calculated using the share of fixed costs absorbed by each value chain present at the site to weigh the value chain-specific scores. For site strategy discussions a disaggregated analysis of the fit between business strategy and site strategy is especially helpful (see also the portfolio graphic in the case example below).

4.5.3 Lessons Learned from Application Case Studies

Selection of an Individual Production Site

One case study conducted concerned the consolidation of three relatively small facilities within a single country. As no changes to the global value chain were intended but solely a restructuring within the country, no network design phase was required. In a first step four alternative restructuring options were developed. These alternatives did not yet require a location decision but instead focused solely on selecting the target structure of the country's production network (i.e., number of remaining sites and distribution of activities across sites). An AHP model was constructed that contained both qualitative and quantitative criteria such as expected restructuring costs and annual savings potential based on rough estimates. Together with a team of local experts a workshop was conducted to compare the restructuring alternatives and assess the objectives' weights. Contrary to the original beliefs of the local project team, a consolidation into a single site turned out to be the favored alternative by a large margin. Sensitivity analyses confirmed that the results were very stable for a wide range of weights.

In a second phase, alternative sites to locate the consolidated site were evaluated. Four sites were considered with one of the alternatives involving a co-location at an existing site and the others requiring the construction of a new site. To reflect the nature of the decision problem the objective hierarchy described above was modified to include the net present value of the alternatives in addition to the qualitative criteria. On this basis trade-offs between qualitative and quantitative objectives were analyzed. In the case example it turned out that co-locating the plant at the existing site is both the most attractive option financially (mainly because no property rent or acquisition was required) and from a qualitative perspective (mainly due to proximity of the existing site to major customers). It should be noted that no major utilities infrastructure was required for the value chain considered and hence constructing a new site could not be ruled out up front for financial reasons.

Site Portfolio Ranking

The site ranking model discussed above was implemented in the course of the research project at the industrial partner. It is integrated into a regular site strategy and controlling process to cluster the entire site portfolio into different categories and define development tracks for each site accordingly. From the perspective of the company, the main advantages of the model are the use of uniformly defined, fact-based criteria to arrive at a joint rating of each site while at the same time incorporating the peculiarities of the value chains present at a site. The structure of the site ranking model integrates site strategy and business strategy and also allows for more in-depths analyses. For example, the fit between business strategy and site strategy can be analyzed using visualizations like the portfolio graphic of a country's site network shown in Figure 33.

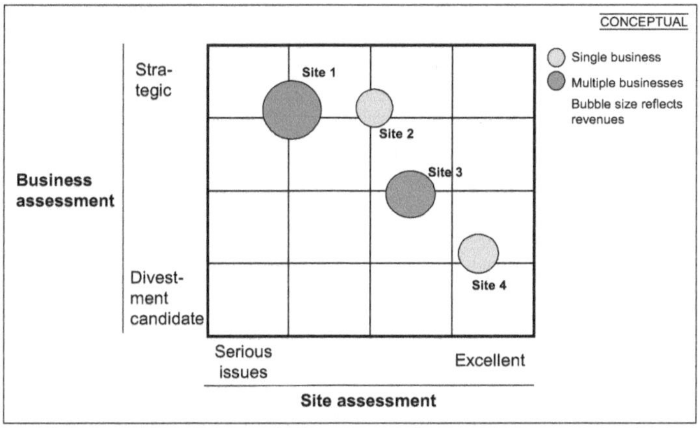

Fig. 33. Example of country site portfolio matrix

The objective weights obtained in the context of the site ranking also illustrate the relative importance of the various location factors for specialty chemicals industry as shown in Figure 34. However, these priorities depend on the type of business considered, the decision context, the scale used and the fact that minimum requirements are assumed to be achieved by all sites included in the analysis. In the application cases, cost performance was considered to be approximately twice as important as qualitative aspects. Within the set of qualitative factors development potential, production know-how and utility availability clearly dominated the site assessment.

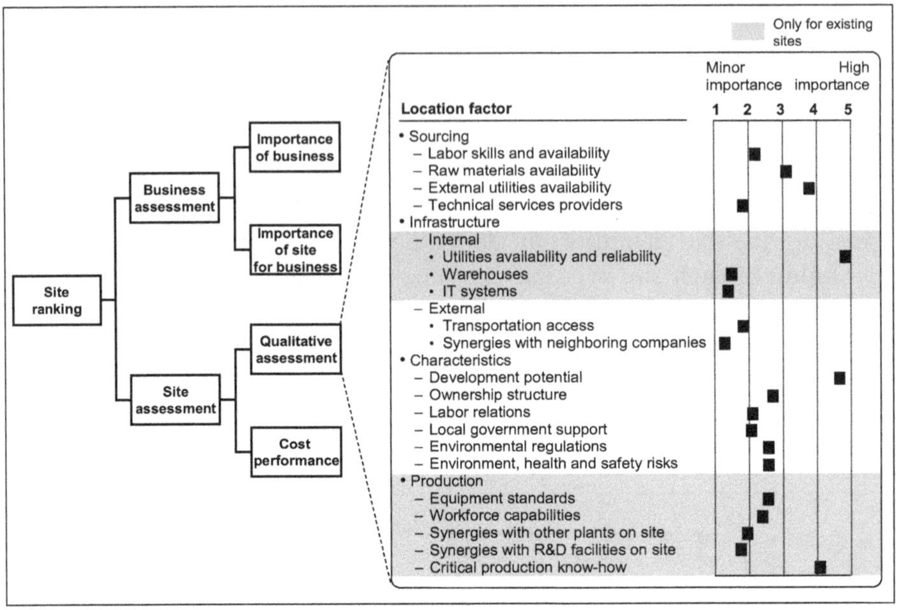

Fig. 34. Relative importance of selected location factors for specialty chemicals

5 Case Study Production Network Optimization

Modeling production systems from industry initially appears to be a straightforward exercise once the model to be used has been customized to reflect the peculiarities of the industry considered. Unfortunately, as also pointed out by many authors such as Pooley (1994, p. 116), Klingman et al. (1986, p. 11) or Brown et al. (1986, p. 227), collecting, preparing and validating the required data is often the greatest challenge. This issue is especially prominent in strategic applications because most of the data has to be forecasted for a long planning horizon and hence cannot be obtained directly from accounting or ERP systems. As also noted by Billington and Davis (1992, p. 587), insufficient data availability is probably a main reason for the reluctance of practitioners from industry to employ optimization models to support strategic planning. The complexity inherent in setting up the data structure for strategic supply network design models is also illustrated by the few publications actually describing the data compilation process (e.g., Geoffrion et al. 1978).

The intention of this case-study chapter is twofold. On the one hand, process steps that were found to be helpful when modeling specialty chemicals production systems are explained to guide practitioners whishing to model their production system. On the other hand, insights that can be gained using a strategic network design model are described based on findings from a number of "key analyses" to convey the value of the model for strategic planning purposes. A pilot application of the model described in Chapter 3, implemented at the industrial cooperation partner, is the basis of this chapter. To maintain confidentiality, the level of detail that can be provided pertaining to the pilot application is limited.

5.1 Developing a Decision Support Tool for Strategic Network Design

5.1.1 Industry Requirements

Properly capturing industry-specific problem characteristics and constraints in the mathematical optimization model can be considered a minimum requirement any decision support tool has to comply with if it is to be used in industry. Additional requirements pertaining to aspects such as user friendliness, data processing and flexibility are important as well. Findings from a survey of facility location software users conducted by Ballou and Masters (1999) indicate that the most important requirements are related to solution optimality, database/data interface, user friendliness and availability of technical support. Commercial systems in use at the time of the survey generally did not fully meet the expectations of the users along those dimensions. The most criticized aspects were related to data input problems, the user interface and a lack of flexibility (however, in somewhat contradictory findings, many users also praised the flexibility of the systems they used). It should be noted that the survey relates to software used for distribution network design. To the author's best knowledge no standard software specifically tailored to the needs of designing complex production networks is available.

The importance of a user interface that fits into the context of the application environment is also stressed by Kallrath (2000, p. 818), whose observations are based on more than a decade worth of application experiences in chemical industry. Breitman and Lucas (1987), reporting an application at General Motors, point out the importance of supporting interactive "what if..." analyses. They found that quantifying the economic consequences of decision alternatives is often perceived to be much more important than obtaining an optimal solution to the unconstrained problem. Ideally, these analyses can be performed real time in workshop settings and the tool is transparent enough to also enable managers without operations research skills to perform the respective analyses (cf. Henrich 2002, pp. 208, 245).

5.1.2 Structure of the Decision Support Tool

Since no commercially available Advanced Planning System was used and the data required for strategic network optimization could not be directly obtained from ERP systems, a separate strategic planning tool was developed for the pilot application. To simultaneously achieve a sufficient degree of flexibility in adapting the decision support tool to changing requirements and enable users without Operations Research skills to utilize the tool, a visual user interface based on an MS Access database was created to interact with the optimization model. The optimization model was implemented using ILOG OPL Studio 4.2 and the solver CPLEX 10. The Access database provides the optimization software with the data required to perform an optimization run and allows to interactively add or remove restrictions to conduct the above-mentioned "what if..." analyses. In order to be able to easily compare the results obtained for alternative calculation runs, a scenario-manager is integrated into the database. The optimization software returns the results, including detailed cost data, obtained via additional decision variables integrated into the model for this purpose, to the database. The user can then evaluate the solution based on standardized reports and/or spreadsheet analyses.

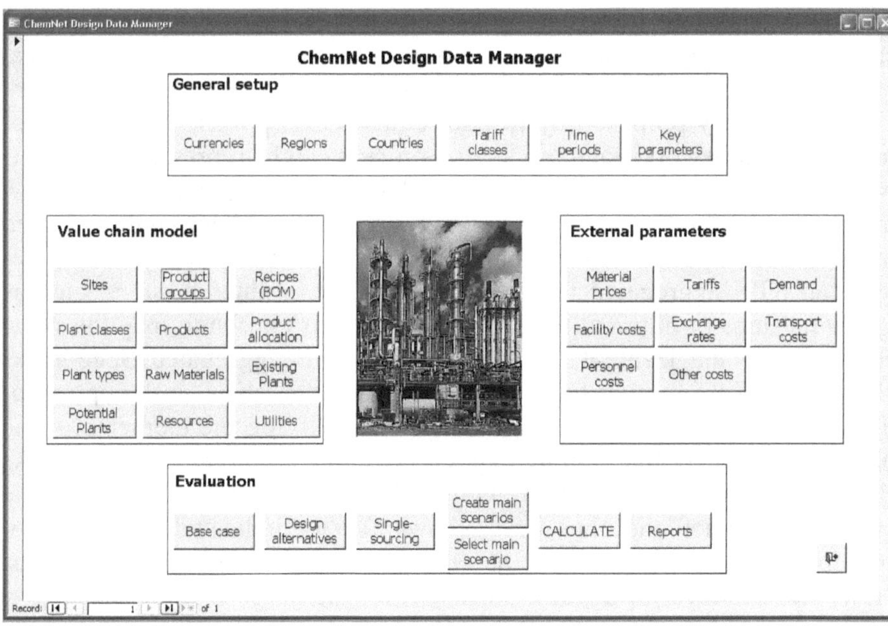

Fig. 35. Main user interface of network design tool

The visual user interface is clustered into four main segments (cf. Fig. 35): *general setup, value chain model, external parameters* and *evaluation.* The general setup items and a subset of the external parameters (e.g., transportation costs, exchange rates) can be used across multiple value chains whereas most other items need to be created separately for each value chain analyzed. Key steps required to properly model a value chain and generate the parameter forecasts/scenarios and core analyses that can be performed based on the model are discussed below. While the software developed in the course of the research project still has to be considered a prototype implementation, it was found to be helpful throughout the analysis of the real-world production network of the industrial partner.

5.2 Creating the Value Chain Model

5.2.1 Mapping the Current Value Chain Configuration

At the beginning of any production network design project a joint understanding of the value chain considered and the main elements belonging to this value chain has to be established. To this end it is helpful to graphically map the value chain. A common form of representation for chemical production processes that can be used to do so is the State-Task-Network (STN) originally proposed by Kondili et al. (1993, pp. 213-215). According to this framework, each production process can be represented as a network of the process steps (tasks) required to produce a product and the raw materials, intermediate products and finial products (states) used or created in the course of each production step. Here, to eliminate unnecessary details, an STN aggregated to the level of major production steps such as synthesis steps or physical transformation steps is chosen to map the value chain. While the level of detail covered by this aggregated State-Task-Network is still too high for representing the value chain in the optimization model, it is a good starting point for identifying the elements of the production process that require explicit modeling in strategic planning. As Blömer (1999, p. 30) points out, the STN represents the recipe of the chemical product and thus only allows for an indicated allocation of (grouped) tasks and states to the corresponding production equipments. However, this level of detail is sufficient for the purposes of strategic network design. The STN of the value chain considered for the case study is shown in Figure 36.

In a first step the core value chain has to be defined. Typically, the tasks performed internally at most or all sites constitute this core value chain. Depending on the level of vertical integration, the tasks might be spread across different plant classes. In the case example shown in Figure 36, the core value chain consists of a single plant class. Often, the level of vertical integration differs across the production network. For example, intermediate s2 is purchased at some plants and supplied internally at others. In these cases, the decision of whether or not to include the production of the respective intermediates in the value chain analysis depends on two additional factors:

- The relative importance of the intermediate for the core value chain (which can usually be determined by analyzing the share of material costs originating from the intermediate).
- The share of overall internal capacity for the intermediate allocated to the core value chain (to account for intermediates which are used across various value chains).

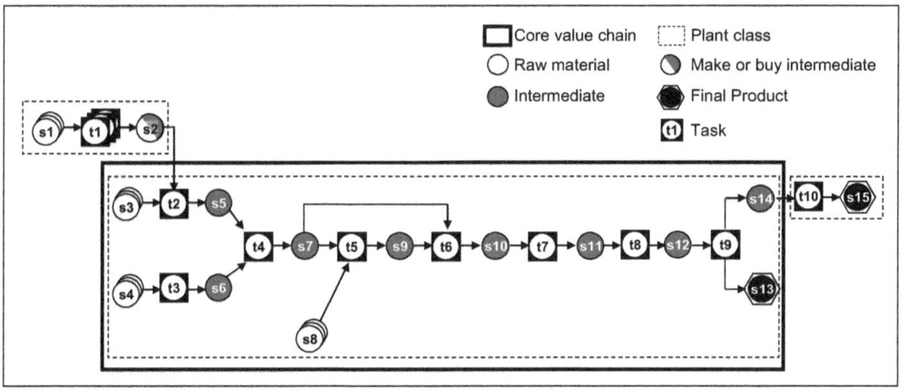

Fig. 36. State-Task-Network for case study example

In the pilot application of the model, s2 was purchased externally at all sites except at the major production site of the value chain. A closure or significant downsizing of this major production site was excluded from the solution space up front. Therefore, it was decided not to include the production of s2 in the model even though the value chain analyzed is the only one using s2 as an intermediate.

Within each plant class to be considered the production process consists of a number of chemical and/or physical tasks. While the type and/or quantity of input factors such as raw materials and utilities might differ significantly between products, the characteristics of the production process within a single plant class are usually similar (except for multi-purpose plants). Therefore, the activities taking place within a single plant class can normally be combined for strategic planning purposes. The overall capacity of a plant is then determined based on the "bottleneck" (usually the major synthesis step, here t4). This bottleneck step is also used to derive the capacity consumption factors for the various products.

In some cases, product mix-dependent effects might have to be considered. For example, only a subset of the products requires processing step t5. Reflecting this fact, the installed capacity of the equipment used for t5 was found to be less than the overall plant capacity at most plants. In situations like these, an additional capacity restriction is required to ensure that only feasible product-plant allocations are proposed by the model. If task t5 requires substantial resources such as dedicated equipments or personnel, the model has to be further extended to explicitly model this resource and the costs associated with its utilization based on the production volumes allocated to the respective plant.

Once the core value chain to be modeled is defined, the existing production network needs to be analyzed. At the outset the architecture of the production network with respect to types and location of plants, current allocation of products and markets and performance levels of the different plants have to be understood (cf. Fig. 37). Also, a joint understanding of the key success factors driving the respective business (e.g., costs, quality, delivery times, etc.) is required. Additionally, transparency is needed regarding planned site restructurings that might affect the value chain in order to integrate these into the analysis. Furthermore, the controlling indicators suggested in Chapter 2.4.5 should be established to get a first idea of the challenges the supply network faces.

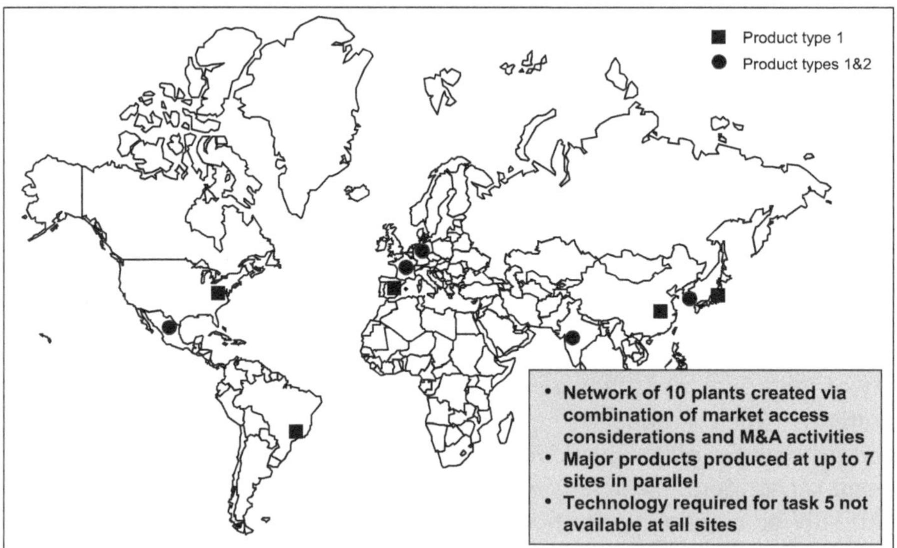

Fig. 37. Current structure of production network

5.2.2 Aggregating Demand and Product Data

As for example Shapiro (2001, pp. 228-232) and Shapiro et al. (1993, p. 110) point out, data aggregation is essential when creating strategic planning models. The most critical aggregation steps relate to products and demand zones considered. Since transportation costs usually constitute a relatively small share of total costs in specialty chemicals, an aggregation of demand at the country level normally suffices for strategic planning purposes. In the pilot application, 15 countries constituted more than 90% of total demand with the largest market representing approximately 40% of total demand, resulting in a relatively small number of demand locations. For very large countries such as the U.S. or China, a split into "subcountries" can be considered if demand is not concentrated at certain areas (e.g., coastal industrialized areas in China) but spread across the entire country (cf. Kouvelis et al. 2004, p. 129).

Specialty chemicals value chains often contain a large number of different products. Considering the data requirements discussed above, it is critical to sufficiently aggregate the product portfolio. In a first step a classical ABC analysis was performed to identify the most important products. The pilot value chain analyzed consisted of a total of 350 different products. As can be seen in Figure 38, the top 10% of products already represent more than 60% of the total production volume. The share of total production volume that should be modeled explicitly depends primarily on the variability of raw material and - to a lesser extent - processing requirements across the product portfolio. In the case study 20% of products representing 80% of demand were modeled explicitly.

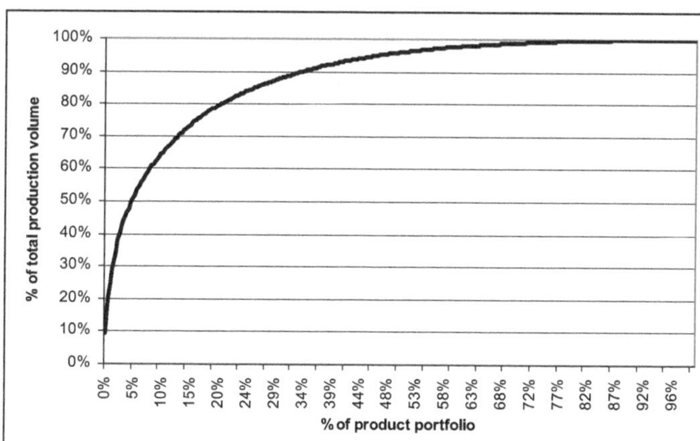

Fig. 38. ABC analysis of product portfolio

Products not modeled explicitly were clustered into groups of similar products according to their raw material consumption profiles. To account for the fact that non-chemical process steps can also constitute a considerable share of the processing activities, within each of the clusters obtained this way, sales price clusters were formed if required. The assumption underlying this approach is that for products with similar raw material consumption sales prices are correlated to the complexity of non-chemical processes, whereas raw material consumption profiles are sufficient to assess the complexity of the chemical processes.

Due to the nature of the product portfolio considered in the pilot application, it was possible to create the clusters rather intuitively without using a formalized methodology such as cluster analysis. When needed clustering techniques such as those used to form manufacturing cells could be adapted to cluster a more complex product portfolio (see Selim et al. (1998) for a review of different clustering techniques). An application of cluster analysis to aggregate demand locations is reported in Müller (1983, pp. 188-190). An intuitive approach similar to the one proposed above was also used by Nickel et al. (2005, pp. 168-170) for an application in steel industry.

5.2.3 Identifying Cost Drivers for Operating Expenditures

The objective of using a mathematical optimization model is to identify network design alternatives that best exploit structural cost differences between various locations and resolve trade-offs between different cost elements such as production cost advantages and additional transportation/tariff costs. Therefore, relationships expressing the costs that will be incurred as a function of cost drivers (decision variables of the model) have to be established. A proper creation of these cost functions is a critical success factor of the overall analysis (cf. Shapiro 2001, p. 234; Vidal and Goetschalckx 1996, p. 13).

Prioritizing Cost Items

The level of aggregation used in strategic planning makes obvious the need to focus on core cost items that actually determine the economic viability of alternative network designs. A two-step approach was used to prioritize cost items.

In a first step the cost structure of a product representative for the value chain was analyzed both for a high-cost and a low-cost country (cf. Fig. 39). As a result, it was decided to explicitly model material costs, labor costs, energy costs, duties and transportation costs. A more detailed analysis of the raw material costs led to the conclusion that usually three or four raw materials are responsible for more than 90% of total raw material costs allowing for a substantial simplification of the modeling requirements. Additionally, plant fixed costs were included in the model to account for repair and maintenance costs, plant management and ecology costs.

In a second step, for the cost elements within the prioritized categories, location-dependent price and availability differences were analyzed. The intention of this second step was to exclude all cost drivers not showing a location dependency from further analysis. A comprehensive study conducted by KPMG (2006) in developed countries found that labor costs are the most location-sensitive cost category, but great differences were also found in property costs and energy costs. In the application case, mainly due to the effects of tariff structures, even for commodity raw materials with a world market price denominated in a single currency an explicit modeling was deemed to be needed and hence no further simplifications were achieved.

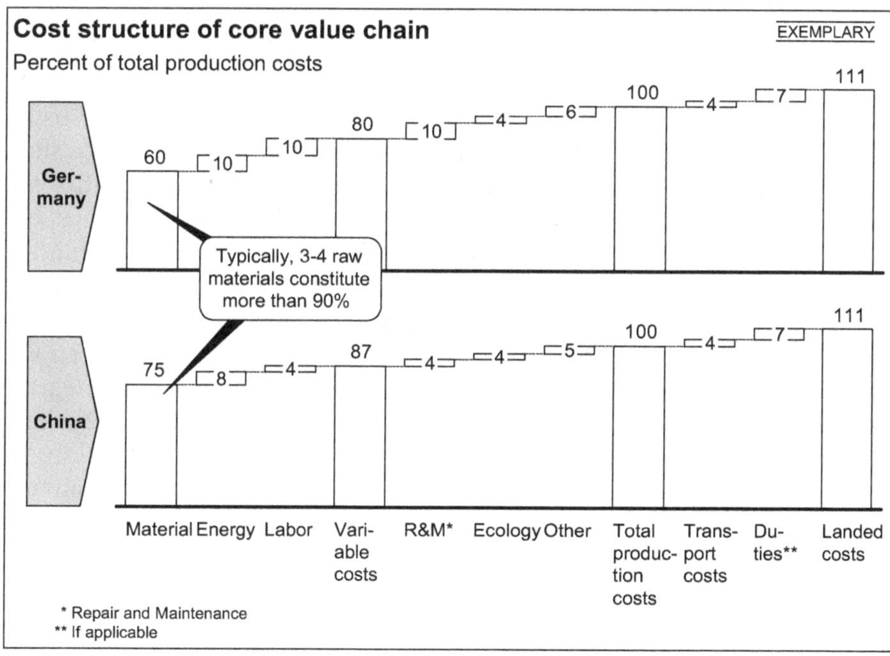

Fig. 39. Cost structure of core value chain

Developing Cost Functions

Shapiro (2001, p. 246) distinguishes between four types of cost data required to parameterize supply network models: product costs, process costs, facility resources costs and facility overhead costs. As Shapiro (1999, p. 297) points out, this cost driver structure is very similar to the structure underlying Activity Based Costing (ABC). ABC was originally proposed by Cooper and Kaplan who use the terminology of unit-level activities, batch-level activities, product-sustaining activities, customer-sustaining activities, product-line sustaining activities and facility-sustaining activities (cf. Cooper and Kaplan 1998, p. 212). In the context of chemical (batch) production systems these could be restated as

- *Unit-related costs*: e.g., raw materials, energy, quantity-proportional production operations such as milling or packaging operations
- *Batch-related costs*: e.g., setup and cleaning but also production processes not proportional to quantities due to batch operation
- *Product-related costs*: e.g., costs for holding specific raw materials or using dedicated equipment
- *Facility-related costs*: e.g., equipment maintenance, plant management

Due to the similarities between ABC and the requirements of supply network optimization models, cost data would ideally be derived using ABC. In such a costing system for each cost element cost drivers are established and costs are allocated to products to the extent that they utilize the respective cost drivers. ABC typically assumes linear relationships between the cost drivers and the cost items because it is employed within the boundaries of the existing production system (cf. Shapiro 2001, p. 249). Accordingly, ABC might have to be extended to incorporate the effects of discontinuities such as capacity reductions, etc., which characterize strategic network redesigns. A more detailed discussion of how to integrate ABC and OR models can be found in Shapiro (1999, pp. 302-309) where industry examples are provided as well. For comprehensive discussions of ABC and guidelines for implementation of such a costing system the reader is referred to Cokins (2001) or Hicks (1999) and for the theoretical foundations to Cooper and Kaplan (1998).

Unfortunately, most companies, including the industrial cooperation partner, do not use ABC. Instead, data had to be obtained from standard accounting systems. Cost functions linking each cost item to decision variables of the network design model had to be created both for existing and potential product-plant combinations. To do so, cost items were grouped into unit-related, batch-related, production line-related, plant-related and site-related costs and the cost functions described in Table 22 were established. The cost function to be used for direct labor depends on the type of plant considered. In batch plants often the number of batches and not the production volume (processing time adjusted or not) determines the personnel required. At continuous plants and highly automated batch plants the tasks performed by operators tend to be of a supervisory nature and hence the capacity consumption of a product determines the personnel requirements. For transportation costs full truck/container loads and only one mode of transportation were assumed resulting in a linear function. While an allocation of indirect costs to individual products is not required for the network design model to fulfill its purpose, this was deemed important by management to assess the results of network design changes. To accommodate this request, allocation rules for an ex post calculation of unit costs were defined for indirect cost items, too.

Table 22. Cost drivers underlying model cost functions

Cost item	Level	Cost driver	Cost function	Unit cost calculation
Material costs	Unit	Production volume	Linear	
Energy costs	Unit/ plant	Production volume/plant operational	Linear	
Transport costs	Unit	Transportation volume	Linear	
Import duties	Unit	Import volume	Linear	
Direct labor (operators)	Unit/ batch	Production volume (adjusted for capacity requirements) or number of batches	Linear	Average batch size if number of batches used
Repair & Maintenance	Production line	Number of operational production lines	Step-wise fixed	Utilization of technical capacity
Ecology	Unit/site	Production volume/ site operational	Only variable share included except for single value chain site	Site-specific surcharge on material costs for fixed share
Other fixed costs	Plant	Plant operational	fixed	Utilization of equipment capacity
Site fixed costs	Site	Site operational	Only included for single value chain site	Not part of unit costs

To further complicate matters, the split of cost items into fixed and variable costs is not consistent across sites. For example, wastewater treatment costs are primarily fixed costs in cases where the company operates its own treatment plant. If communal facilities are used or the site is located at an industrial park, typically a larger proportion of the costs is linked to actual consumption and thus variable. Similarly, energy supply prices usually contain a fixed component linked to the peak capacity guaranteed or the base-load supply (capacity charge) and the price per unit consumed (unit charge). Depending on the energy mix of a site and the contract situation this can lead to substantial differences between sites.

The decision to primarily rely on linear cost functions can be challenged. However, this simplification is in line with reports from literature indicating that linear functions are sufficiently accurate for strategic planning purposes (cf. Billington and Davis 1992, p. 590). If required, functions such as concave raw material purchasing costs can be implemented using stepwise linear cost functions. Site-related costs were only included if the value chain is the sole user of a site. In all other constellations changes at the value chain level usually do not significantly affect the costs incurred at the site level.

Throughout the data collection phase unexpected complications were encountered in the pilot application because calculations obtained from the ERP system were often not comparable across plants. Even if the same ERP system was used (which was not the case for all sites considered), different conventions were used for distributing costs across products. Additionally, recipes pulled from the ERP system often did not reflect the actual recipes employed in production. For example, significant differences in absolute raw material quantities required and relative relationship between the different raw materials were found. As similar problems are also reported in the literature discussing practical applications of mathematical modeling approaches (e.g., Kallrath 2000, p. 817; Lee and Billington 1995, p. 46), this appears to be the norm rather than a company-specific exception.

Implementing the Cost Functions

To model unit-level items recipes are represented using a *Bill Of Materials* (BOM) structure. The BOM contains the amount of core raw materials, intermediates and energy required to produce the respective product. Additionally, consumption factors for capacity-restricted resources and utilities are included. At the outset of the pilot application it was intended to use a single BOM for every product that is valid throughout the time horizon at all sites capable of producing the product. However, a comparative analysis of recipes from different sites led to the conclusion that yield levels can differ significantly between sites. Therefore, a default BOM, which at the same time serves as the target recipe to be achieved in new plants, is created. Whenever required, site-specific versions can be created to account for different levels of efficiency at the sites currently producing the product. An exemplary site-specific BOM is shown in Figure 40.

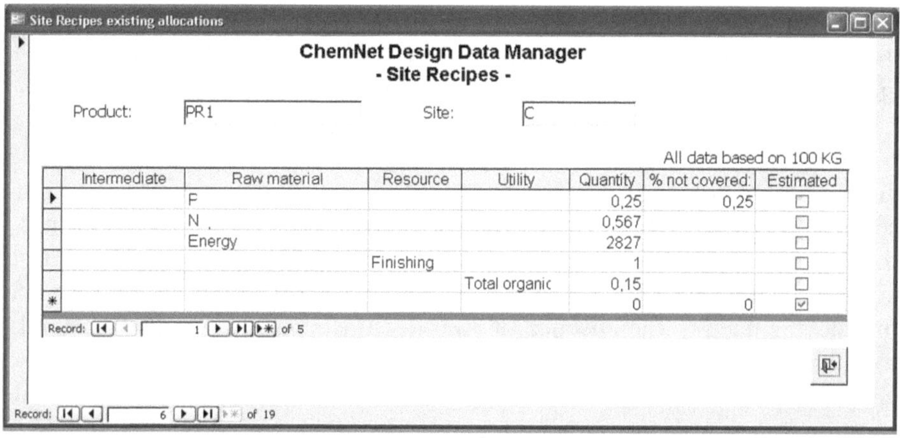

Fig. 40. Bill Of Materials representation in Data Manager

To support the user, step-by-step guiding dialogs are implemented in the database. As long as the site-specific BOM has not been modified by the user, all data entries are marked to be estimates since they are only derived from the default BOM. To accommodate process improvements that are expected over time, the BOM modeling can be further detailed to include time-dependent changes.

Fixed costs are modeled in a similar fashion. For each plant class a default model containing plant fixed costs and production line fixed costs is created. Site specific copies of this default plant are created and the default cost structure is updated based on the actual values observed at the individual sites. The plant model additionally contains the capacity of the plant and the number of employees required to operate a production line. Based on the capacity consumption factors contained in the product description and the number of operators required per capacity unit, variable personnel costs are allocated to the products. If required, variable costs associated with additional resources contained in the BOM are allocated to the products in identical fashion.

Validating the Model

User acceptance of the results obtained is often the key barrier to successfully employing operations research methods in strategic planning (cf. Pooley 1994, p. 120). As amongst others Bramel and Simchi-Levi (1997, p. 262) and Billington and Davis (1992, p. 592) point out, a validation of the cost functions integrated into the model is critical to ensure acceptance by top management.

In a first step, the current baseline should be reproduced by restricting the model to replicating the status quo (cf. Kallrath 2000, p. 818; Shapiro et al. 1993, p. 112). The resulting cost structure must be in line with the data available from accounting or controlling systems or it must at least be possible to explain the reasons for deviations. For the proposed model a very close match was achieved for material and energy costs. For personnel costs and repair and maintenance costs the degree of conformity depended on how these costs were allocated to products in the accounting system. In the optimization model the capacity consumption factor of the products is used in the absence of Activity Based Costing data. Especially for personnel costs different results might be obtained if more accurate cost driver data were available. Whenever deviations were caused by allocation rules, validation was achieved by demonstrating that the total amount of costs incurred is in line with current data and ensuring a reasonable allocation across the product portfolio. Additionally, it should be noted that due to the cash flow – based approach no depreciation costs are included in the model.

In a second step, to further increase acceptance of the model, "what if..." analyses, often based on extreme scenarios or significant modifications of a single parameter, were used to generate a joint understanding of how the model reacts to parameter changes.

5.2.4 Identifying Alternative Value Chain Configuration Options

Investment Opportunities

Optimization models are capable of evaluating all possible plant-site combinations simultaneously. However, this approach would considerably increase the data preparation efforts required since for each plant-site combination both investment and operating expenditures would have to be estimated. Additionally, calculation times increase significantly with the number of alternative investment opportunities.

In practice, in a first step potentially attractive investment countries were chosen. To do so, management ranked countries according to their perceived attractiveness for the value chain considered. Depending on the type of value chain and the corresponding competitive priorities, the ranking might be driven by aspects ranging from cost considerations via availability of certain resources to proximity to customers. For the countries selected by management investment options were integrated into the model.

Whenever possible, the results of the qualitative ranking of the current site portfolio (cf. Chap. 4) were used to identify existing sites located in or close to the countries favored by management to generate investment alternatives that are in line with the overall site strategy.

For each potential site the relevant subset of plant types to be considered was defined by ruling out some of the size/degree of automation combinations theoretically possible. While some combinations are infeasible already at the size/degree of automation level (e.g., small, highly automated or large with low automation, cf. Chap. 3.3.3), adding the country level helps to further reduce the number of potential combinations. Typically, in high-cost countries neither small plants nor plants with a low level of automation are considered whereas in low-cost countries high degrees of automation are usually avoided.

Network Design Principle

Changes in the competitive landscape might require the analysis of alternative value chain configurations. An initial understanding of viable configuration alternatives can often be gathered by analyzing the production networks of competitors. However, to truly understand the options theoretically available and the ones that could actually be implemented in practice two core analyses were found to be helpful.

At an aggregated level migrating the network in the direction of one of the generic network design principles proposed by Schmenner (1979) or defining different strategic roles for each plant according to the plant roles suggested by Ferdows (1989) can be considered (cf. Chap. 2.1.4). For example, the economic effects of producing the top 20% of products in every region (market area plants) versus focusing each plant on a certain subsegment of the product portfolio (product plants) or transferring all non-innovative products to low-cost countries could be assessed. The motivations behind pondering whether to migrate an existing network in one of those directions are usually not purely of quantitative nature but extend into overall manufacturing strategy. The core advantage of having a production network design model available is that it allows to quantify the financial implications of alternative network design strategies.

At a more disaggregated level a deconstruction of the core value chain can be considered (for example to focus certain plants on specific process steps). In a first step all technically separable elements of the value chain should be identified and their respective processing characteristics established. To illustrate the concept behind this analysis, the State-Task-Network of the pilot application value chain depicted in Figure 36 is revisited in Figure 41. For example, after completion of t7 the product is dried and can be stored/transported without limitations. Hence, tasks t8 (a granulation process) and t9 (a blending and packaging step) could be separated from the core value chain.

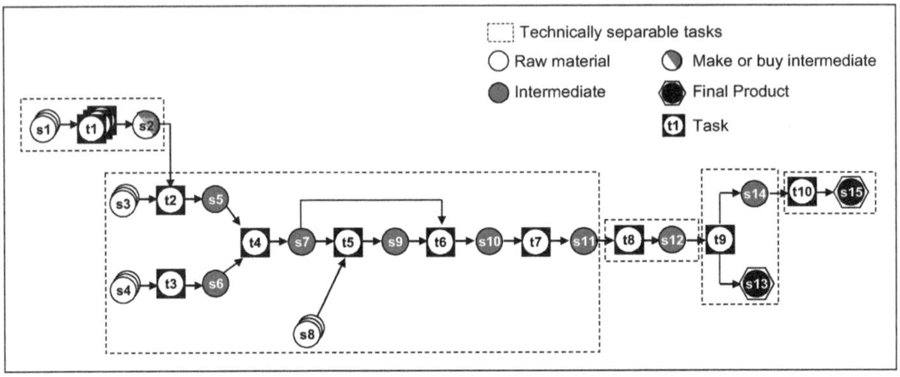

Fig. 41. Disaggregation of STN into technically separable elements

Once the technically separable groups of tasks are identified, the economic feasibility of actually reconfiguring the value chain in a different way can be analyzed. For example, the blending and packaging task t9 might not only homogenize the product but create multiple product variants via blending of basic products available after t8. In the majority of specialty chemicals value chains, the predominant mode of production is make-to-stock. If the lead time characteristics of t9 as compared to customer requirements allow this, shipping products in state s12 in a bulk container from producing sites to major markets and setting up "blend & pack to order" facilities might be an option. Alternatively, if the packaging process is labor intensive, bulk shipments can be considered for those products supplied from high labor cost countries to low labor cost countries. In these countries a packaging operation could be set up at an existing production site (if available) to perform this last, labor-intensive process step. More generally, by restricting the maximum lead time allowed for deliveries to the destination market, the economic consequences of switching certain steps to a make-to-order system can be assessed. The rationale behind these considerations is the same as the design and postponement concepts for example described by Lee and Billington (1995, pp. 53-55) for electronics supply chains at Hewlett Packard.

5.3 Establishing and Forecasting External Parameters

5.3.1 General Considerations

The parameters that have to be established for all existing and potential plants and products include amongst others input factor costs, demand volumes and prices. While some of these parameters such as plant operation costs or transportation costs can be determined (and to a lesser extent forecasted) relatively easily, others such as plant construction costs or demand volumes and prices are much more difficult to establish.

A critical aspect that has to be kept in mind is that when generating these forecasts, inflation has to be treated consistently throughout the model. Since different input factors show drastically different inflation rates (e.g., significant personnel cost increases in developing countries), inflation should be considered explicitly in the model. Doing so is especially complex in countries with high inflation because of the time lags usually observed between rises in input factor costs and restoration of profit margins to normal levels (cf. Hayes and Wheelwright 1984, pp. 144-146; Shapiro 1978, pp. 14-15). If an after-tax objective function is used, the matter is further complicated by the fact that items such as tax benefits from asset depreciation are based on historical purchase prices whereas tax-deductible interest payments (may) also depend on expected inflation rates (cf. Freidenfelds 1981, pp. 43-48).

A detailed discussion of forecasting techniques is not in the focus of this work (the reader is instead referred to Chase et al. (2006, pp. 510-557), Günther and Tempelmeier (2005, pp. 142-150), Heizer and Render (2005, pp. 81-118) or Shapiro (2001, pp. 257-261) for overviews on the subject, Hanke and Wichern (2005) for a detailed presentation of different techniques and Haehling von Lanzenauer and Sprung (1982) for a scenario-based approach to forecasting future levels of inflation). However, below a few key aspects of parameterizing the model are discussed.

5.3.2 Investment Expenditures

Investment expenditures have to be estimated both for expansions of existing plants and for construction of new plants. However, a comprehensive engineering study is too costly to be performed as part of an initial strategic analysis. Instead, estimation techniques combining historical data with findings from empirical research can be used.

Gallagher (1967) discusses different techniques for rapid estimation of plant construction costs in chemical industry. One technique is to determine the costs of major equipments, which can usually be obtained from vendors, and use so-called *Lang Factors* derived from statistical analysis of a large number of projects to estimate overall investment expenditures. As these factors depend on the characteristics of the equipment (e.g., carbon-steel vs. stainless steel), they have to be applied with great care. If data from recent investment projects into the same type of technology is available, it is also possible to derive estimates based on this data, inflation rates and published capacity coefficients. Exponential coefficients, derived from statistical analysis of large numbers of projects, allow to calculate costs for different capacities based on the known costs for a certain capacity. The coefficients are available for many different plant classes and equipments. For example Remer and Chai (1990) and Remer and Chai (1993b) provide coefficients at the equipment level for chemical industry and Remer and Idrovo (1993) provide data for pharmaceutical and biotechnology equipment. Data at the plant/production process level can be found in Remer and Chai (1993a).

Once the investment expenditures for plants/production lines are established for a single country, they have to be localized to account for cost differentials between countries. To this end, location adjustment factors can be employed (cf. McMillan and Humphreys 1990). Location factors for chemical industry can for example be obtained from SRI Consulting. A comprehensive overview of commercial sources for cost, inflation and location factors is published in Remer and Mattos (2003). Kohn et al. (1997) discuss how to construct country-specific factors from the U.S.-based *Chemical Engineering's Plant Cost* index.

5.3.3 Transportation Costs

The level of detail required when modeling logistics costs depends on the relative importance of logistics costs for the supply network considered. In industries where the share of logistics costs is high (e.g., > 10% for aircraft components or up to 50% in petroleum industry from oil field to final market) a detailed modeling of the various cost items is required (cf. Zeng 2003, pp. 272-273). For the application areas considered in the course of this work an aggregated approach covering average "from country – to country" costs was found to be sufficient due to the limited importance of transportation costs. With respect to forecasting future rates, it is questionable whether the trend of decreasing costs for intercontinental freight rates observed throughout the last years (cf. Meyer 2006, pp. 72-76) will continue in light of rising energy prices. Also, asymmetries in global freight volumes lead to prices that are route-dependent. For example freight rates for transports originating from China are approximately three times higher than those for transports to Chinese destinations.

An interesting level of complexity is added if the network design model is required to provide precise data on currency exposure. As shown in Figure 42 various currencies can be involved in a simple transport of a product from a site to another site or a market.

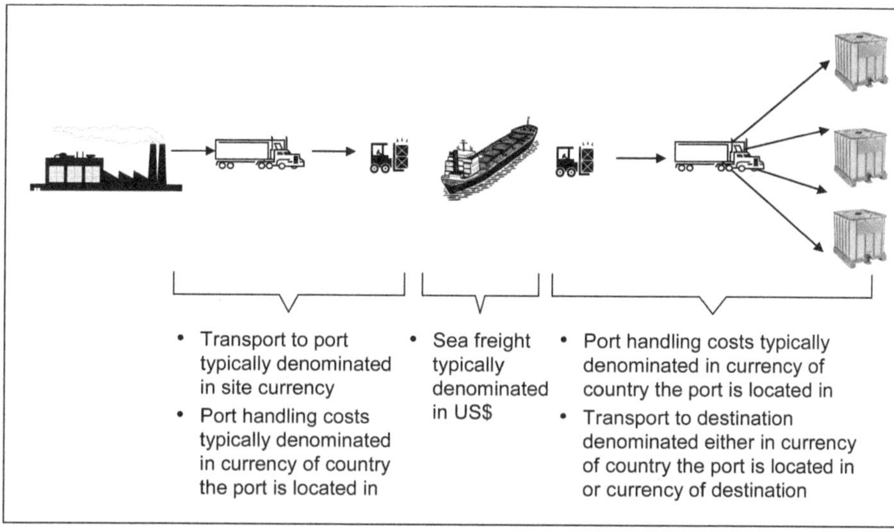

Fig. 42. Currency distribution in transportation costs

5.3.4 Personnel Costs

Since global specialty chemicals companies already operate sites in most relevant countries, average personnel costs per full-time employee can be established relatively easily. However, location-specific adjustments of the number of operators required to run a production line/plant are required. Firstly, different degrees of automation are usually deployed in developed and developing countries. This aspect is to some extent included in the plant type definition via personnel requirements that depend on the degree of automation. However, even for technically identical plants the number of operators required per shift usually differs between developed and less developed countries due to different levels of personnel productivity. Secondly, annual working hours and consequently the relationship between operators required per shift and total number of operators required differs strongly between countries. For example, at the lower end of the spectrum workers in France work on average less than 1,600 hours per year whereas in countries such as India, Mexico or South Korea the average annual working hours exceed 2,200 hours. Figure 43 contains data for a number of countries to further illustrate the importance of this aspect. The data reported is in line with the results of a study conducted by UBS (cf. UBS 2003, pp. 21-25).

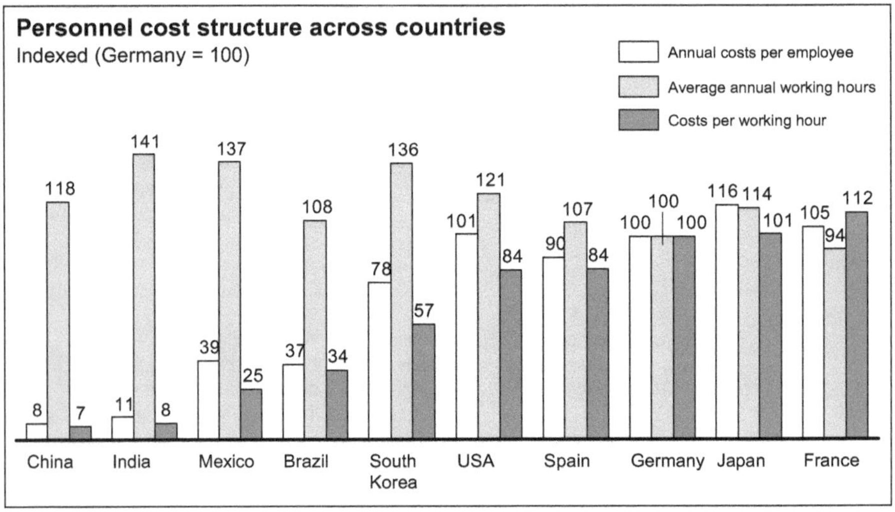

Fig. 43. Personnel cost structure (own analysis based on 2005 company data)

In the past, labor cost advantages have often been proven to be of transitory nature. For example, average annual real personnel cost increases of more than 10% were observed in Taiwan in the years between 1975 and 1990 (cf. Bürklin 1993, p. 106). This has forced companies whose strategy it is to always follow lowest labor costs to relocate to ever more "remote" locations (cf. Schmenner 1997, p. 112). A prime example is the textiles and clothing industry, that – within the restrictions imposed by import quota systems that were in place until the end of 2004 – consistently chases lowest labor costs because here labor costs are by far the most important cost item (cf. Nordas 2004; Hinterhuber et al. 1994). Since the share of labor costs is relatively small in chemicals the forecast quality is not as critical as in other industries However, this type of development should be kept in mind when forecasting labor costs for developing countries.

5.3.5 Exchange Rates

A key problem associated with modeling exchange rate effects is that it is practically impossible to forecast exchange rate movements (cf. Kasa 1995). The approaches used in literature vary in their degree of sophistication. Simple approaches such as the ones used by Bhutta et al. (2003, p. 207) and Mohamed (1999, p. 85) are based on linear trend functions. Huchzermeier and Cohen (1996, pp. 105-107) propose a multinomial approximation method using risk-free interest rates, variances of exchange rate movements and the correlation across countries to generate exchange rate scenarios and transition probabilities. Lowe et al. (2002, p. 579) propose to generate exchange rate scenarios using either the degree of fluctuations observed historically from one year to another or ad-hoc currency scenarios (e.g., +/- 30%). For long-term forecasts explicit scenarios are generated for three time periods based on historical patterns observed over three-year intervals. The average of the three-year interval explicitly considered is used for the remaining periods. Dufey and Mirus (1981) discuss general issues and options in exchange rate forecasting. In the pilot application, for currencies with significant exposure discrete scenarios derived from historically observed exchange rates were used to assess the sensitivity towards currency fluctuations (cf. Chap. 5.5.2).

5.4 Performing Analyses and Evaluating Results Obtained

As Geoffrion and Powers (1980, pp. 25-30) point out, the benefits of employing strategic network design models by far exceed the calculation of an optimal solution to a given problem. Among the additional uses assessing the impact of alternative environmental scenarios, evaluating proposed business decisions and policies and the sensitivity of the preferred network design to variations of external factors feature most prominently. The integration of these aspects into the decision support tool is discussed below considering both, the creation of the required data and the implementation into the optimization model. Also, sample evaluation reports are presented to illustrate the nature of the insights that can be obtained using the model.

5.4.1 Assessing Alternative Scenarios

Use of Scenario Analysis in Strategic Planning

Scenario planning is a technique that utilizes "pen-pictures of a range of plausible futures" (cf. Wright and Goodwin 1999, p. 318) instead of a single forecast value to support strategic planning. Scenario planning assumes that it is only possible to identify critical uncertainties and plan for a range of futures that could unfold. Typically, scenarios describe potential future states and possible migration paths that could lead from the current state to the different future states based on a complex network of influencing factors (cf. Hahn and Hungenberg 2001, p. 127). The principle underlying scenario planning is illustrated in Figure 44.

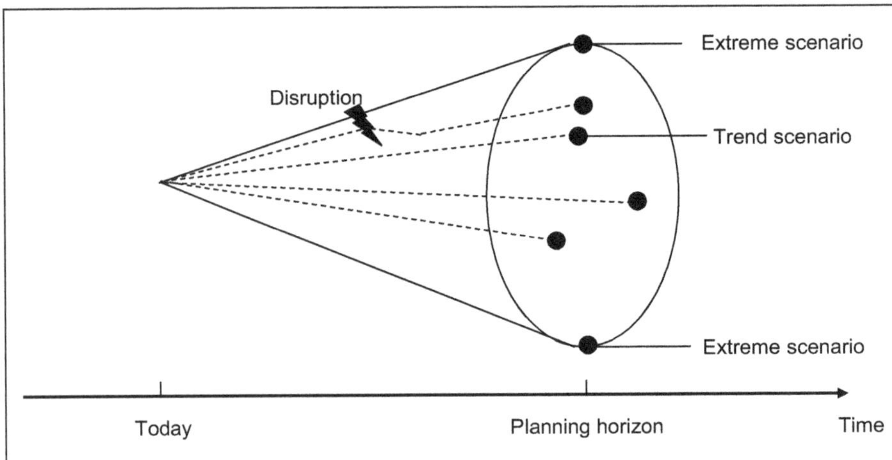

Fig. 44. Principle of scenario planning (cf. Geschka 1999, p. 522)

While it is generally recommended to use scenario planning as a precursor to quantitative decision analysis (cf. Wright and Goodwin 1999, pp. 318-319; Geoffrion 1978, pp. 166-167), Herzhoff (2004, pp. 150-195) found in his empirical study on the use of scenario planning in chemical industry that only 43% of the participating companies actually used scenario techniques.

Generating Alternative Scenarios for External Parameters

The issue of scenario generation can be looked at from a content and a process perspective. The content perspective deals with the question of how to derive meaningful scenarios for individual parameters such as demand volumes or prices. Since an extensive discussion of the various techniques that can be employed to facilitate scenario generation content-wise is not the focus of this work, the reader is instead referred to the relevant literature (e.g., Geschka 1999; Godet and Roubelat 1996; Schoemaker 1995; Schoemaker 1993; Schnaars 1987; Vanston et al. 1977). Herzhoff (2004) evaluates commercial software tools available to support scenario analyses based on a case study from chemical industry. Case histories on the use of scenario planning are for example reported in Moyer (1996), Mobasheri et al. (1989) and Klein and Linneman (1981).

The process perspective deals with the question of how to implement scenario analyses in the optimization model and how to generate the data sets representing the scenarios. Considering the complexity of the data structure underlying production network design models and the large number of data elements that have to be modified, the decision support tool has to provide features to assist the user in generating scenarios for key parameters. Figure 45 shows the features provided to generate alternative demand scenarios. The user can define growth rates both for volumes and prices either at the region/product group level or at the country/product level and automatically generate the corresponding demand data. If required, this automatically generated data set can be further edited manually. Similar scenario generation capabilities can be used to generate scenarios for other relevant parameters.

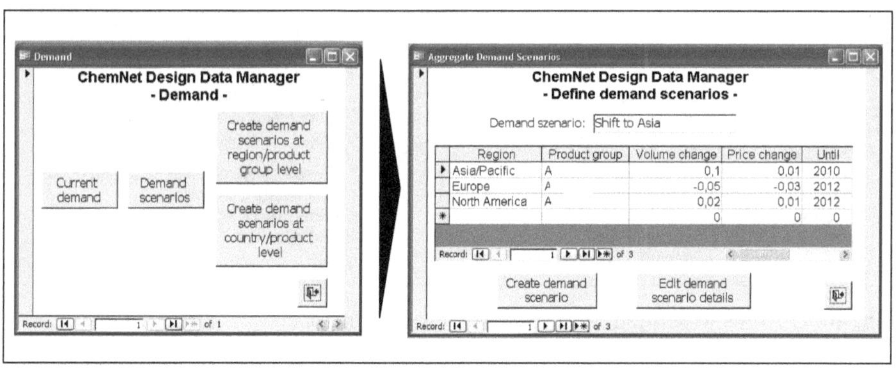

Fig. 45. User support features for scenario generation

5.4.2 Analyzing Network Configuration Alternatives

Performing "What if..." - Analyses in Strategic Planning

Strategic network design models allow for analyses ranging from a completely new (greenfield) design of the value chain to an optimization within the limitations of existing capacities - effectively reducing the model to a tactical one. While many models in literature do not explicitly analyze the restructuring of existing networks, this is in practice the most relevant application area of network design models. A possible approach towards exploring the solution space is presented in Figure 46. By gradually restricting the solution space, decision makers can assess the consequences of operating within an existing network instead of a greenfield situation and attach "price tags" to restrictions imposed by non-quantitative constraints.

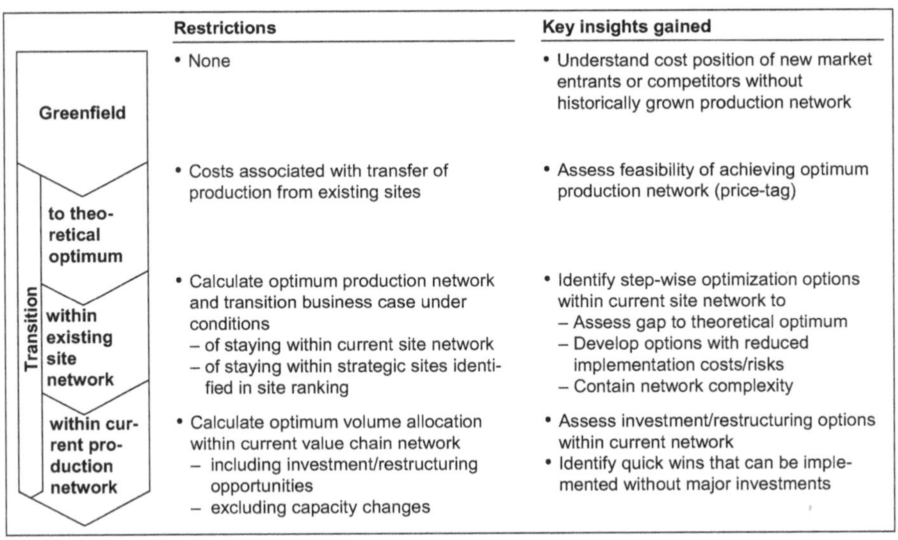

Fig. 46. Options to sequentially restrict the solution space

In addition to this general approach to exploring the solution space, the model can also be used to evaluate individual configuration options such as plant expansions, construction of new plants or closure of existing plants. For questions like "what happens if we close a production line at site X" the model has to provide recommendations on where to relocate affected products and calculate the economic impact of the decision. Depending on the critical success factors of the business ("order winning criteria") additional restrictions derived from manufacturing strategy might have to be imposed on the network design model (cf. Berry et al. 1995). For example, in a make-to-order environment the maximum delivery time could be a critical success factor that has to be restricted.

Creating Alternative Configuration Options

The capability to interactively add or remove restrictions on the production network design without operations research skills is important for the acceptance of the decision support tool. Therefore, several features were integrated into the prototype to facilitate this kind of analysis (cf. Fig. 47). The tool allows to create multiple sets of configuration alternatives that can later be combined into scenarios for evaluation runs (cf. Chap. 5.4.3).

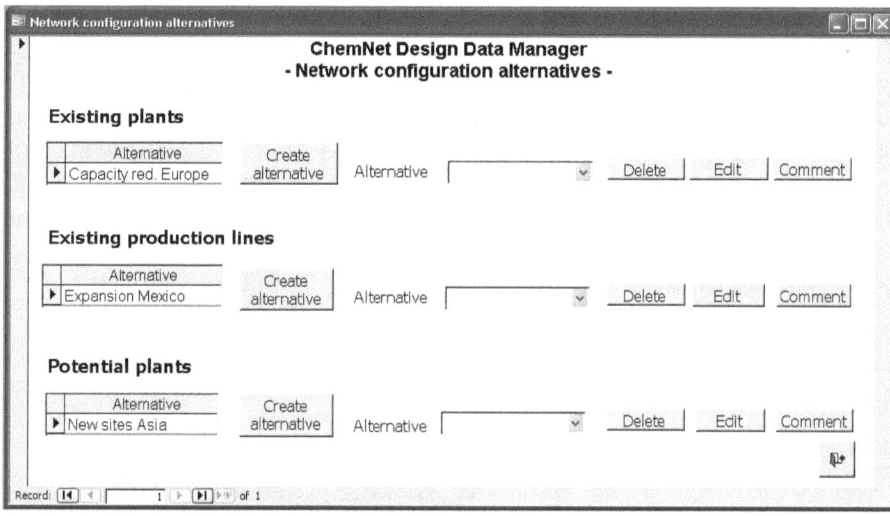

Fig. 47. Managing configuration alternatives

At the structural level additional restrictions can be added that force the model to close or downsize existing plants at a certain time period or rule out a plant closure or downsizing for the respective time periods. Analogous, planned investments can be analyzed by requiring the model to either increase the capacity of an existing plant in a certain time period or open a new plant at a specific site in the respective time period. Managerial constraints can also be applied to other decision types. For example, the total number of product transfers occurring in a given time period can be restricted to account for limited capacity of the experts required to facilitate the product transfers. Similarly, the number of plant expansion or construction projects can be limited based on the engineering capacity available. Restrictions could also be applied on the number of employees that can be laid off at a site in a single time period to restrict the social implications of network design changes.

At the allocation level product-plant and plant-market allocation decisions can be evaluated. In many cases these decisions take the form of regular single-sourcing restrictions. Alternatively, it might be desired to asses the impact of producing certain products only at a single site that is to be selected by the model or pre-determined by the user to reduce the complexity of the network. As discussed in the context of the numerical performance analysis, the latter type of restriction has to be handled with care because of the strong increase in calculation time that was observed.

5.4.3 Integrating Parameter Scenarios and Configuration Alternatives

In order to integrate parameter scenarios and configuration alternatives and to provide comprehensive evaluation features, a *Scenario Manager* has been included in the prototype decision support tool. The main feature of the Scenario Manager is the capability to combine parameter scenarios and design alternatives into comprehensive *main scenarios* for evaluation runs (cf. Fig. 48). The standardized reports introduced in the next chapter use these main scenarios to compare the results of different evaluation runs.

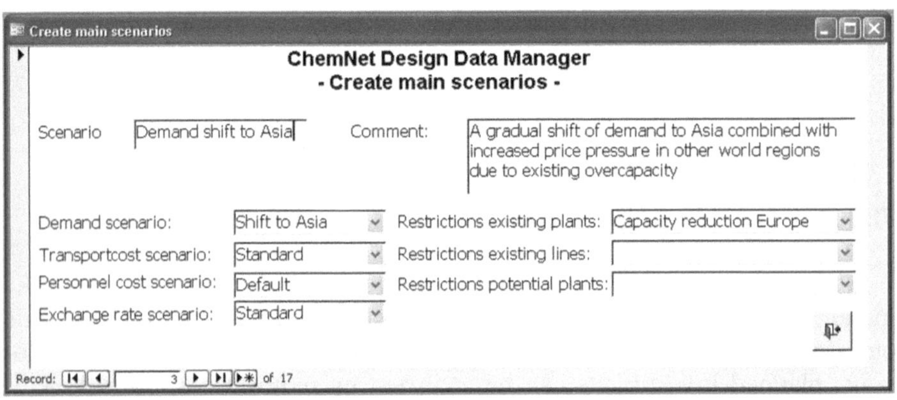

Fig. 48. Creating main scenarios for evaluation runs

Additionally, the Scenario Manager allows the user to create network configuration restrictions from the results obtained in one of the calculation runs by copying the complete network configuration of a solution run into a set of design restrictions. The resulting configuration can then be assessed with respect to its performance under alternative environmental scenarios. This way, an understanding of the solution robustness of the preferred network design alternative can be established.

5.4.4 Standardized Evaluation Reports

In order to evaluate results obtained from different calculation runs, a number of standardized reports were integrated into the decision support tool. The reports can be clustered into the three categories *production network structure, cost structure and investment/restructuring activities* (cf. Fig. 49). Since reports become very complex if they contain data from several scenarios, comparative analyses of different scenarios are performed via data downloads into MS Excel pivot spreadsheets.

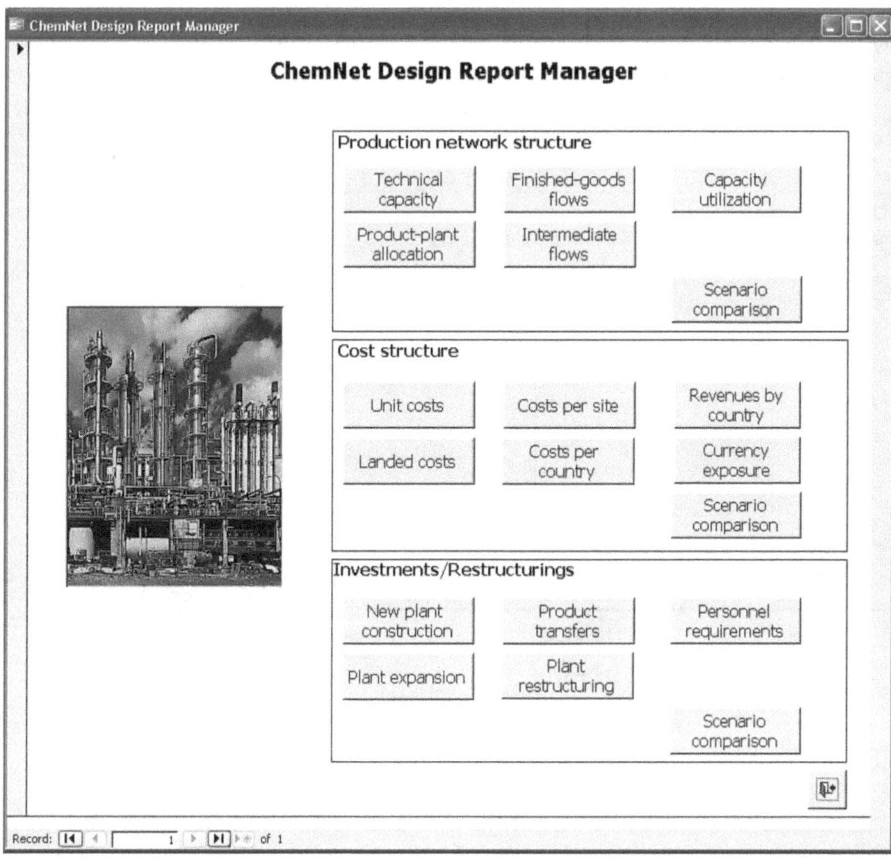

Fig. 49. Report Manager

To illustrate the nature of the reporting features that were integrated into the decision support tool, three particularly interesting/important analyses that can be performed using the optimization model will be briefly presented.

Exchange rate fluctuations can strongly affect the optimal structure of the production network and its profitability. As discussed in Chapter 3.3.5 companies face several obstacles when seeking to properly integrate exchange rate uncertainties into strategic models. Among them the difficulties around creating meaningful exchange rate scenarios and the interaction with other value chains and financial hedging tools feature most prominently. A key analysis that can be performed using the production network design model is to identify the operating exposure of the value chain. Except for the simplifying assumptions used to model currencies in transportation costs, all operating cash flows can be linked to the currency they are denominated in. For investment expenditures a split into locally/globally sourced services/equipments and a definition of the country of origin for globally sourced elements are required to achieve a proper distribution across currencies. Figure 50 shows an exemplary currency distribution for the operating cash flows of a single year.

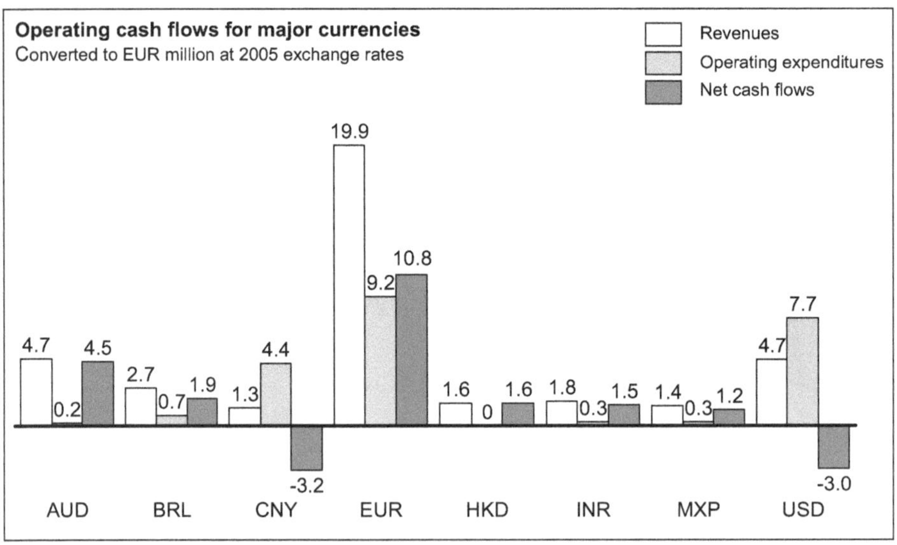

Fig. 50. Operating cash flows by currency

Calculation of landed costs at the demand locations is a particularly important analysis. Traditional analyses often focus on comparing unit manufacturing costs. However, the landed costs determine the profitability and competitiveness of a business. Calculating landed costs cannot be achieved by simply adding transportation costs and import tariffs to (average) unit manufacturing costs. Instead, unit production costs have to be adjusted for duty drawbacks. The following example illustrates the issue:

- A plant in Germany produces a product using raw materials imported from China.
- The import tariff for the raw materials is 6.5% of the raw material value. Hence, assuming 65% raw material costs the import duties on the raw material constitute approximately 4% of total costs.
- If the final product is sold within the European Union these costs have to be included in the unit manufacturing costs.
- If the final product is re-exported, a duty drawback can be obtained and hence the unit production costs have to be adjusted accordingly.

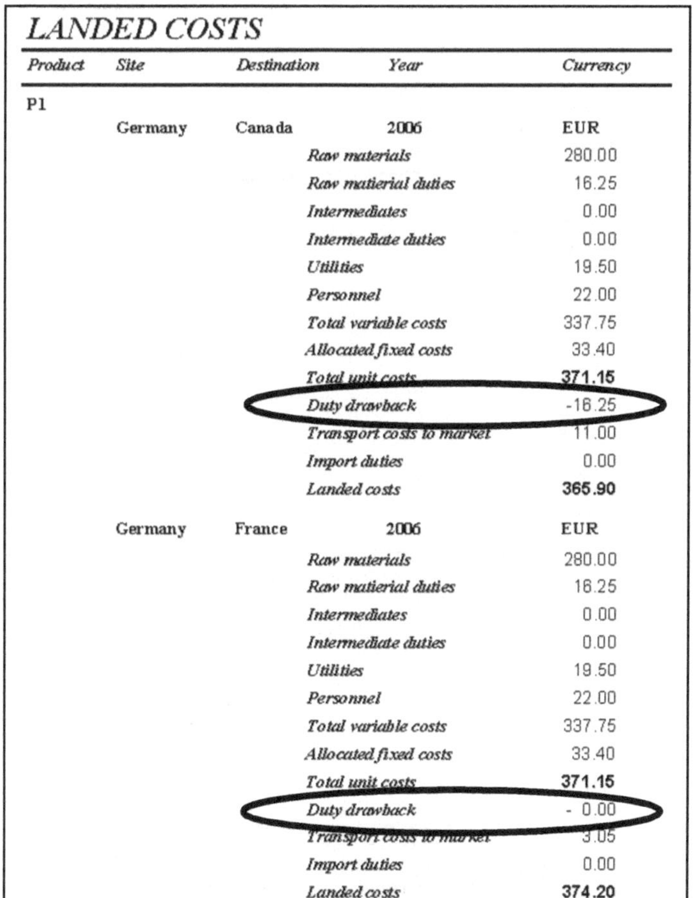

LANDED COSTS

Product	*Site*	*Destination*	*Year*	*Currency*
P1				
	Germany	Canada	2006	EUR
			Raw materials	280.00
			Raw matierial duties	16.25
			Intermediates	0.00
			Intermediate duties	0.00
			Utilities	19.50
			Personnel	22.00
			Total variable costs	337.75
			Allocated fixed costs	33.40
			Total unit costs	**371.15**
			Duty drawback	-16.25
			Transport costs to market	11.00
			Import duties	0.00
			Landed costs	**365.90**
	Germany	France	2006	EUR
			Raw materials	280.00
			Raw matierial duties	16.25
			Intermediates	0.00
			Intermediate duties	0.00
			Utilities	19.50
			Personnel	22.00
			Total variable costs	337.75
			Allocated fixed costs	33.40
			Total unit costs	**371.15**
			Duty drawback	- 0.00
			Transport costs to market	3.05
			Import duties	0.00
			Landed costs	**374.20**

Fig. 51. Landed costs report

The landed cost analysis included in the report manager takes these effects into account, whereas traditional accounting systems often use standard costs or average costs to determine unit production costs irrespective of these considerations. Figure 51 shows an example of the landed costs report.

Another aspect of using the network design model is to be able to compare the results obtained for different scenarios. Using the pivot spreadsheet interface of the database, the user can interactively select the scenarios he wishes to compare and gradually drill down into the available data. The example shown in Figure 52 compares product-plant-market allocations for different currency scenarios. Scenario 1 is based on the current weakness of the US$ in comparison to the Euro, whereas scenario 2 assumes a strengthening of the US$. As can be seen, the exchange rate change leads to a shift of production volumes to the "H" site which is located in the Euro zone.

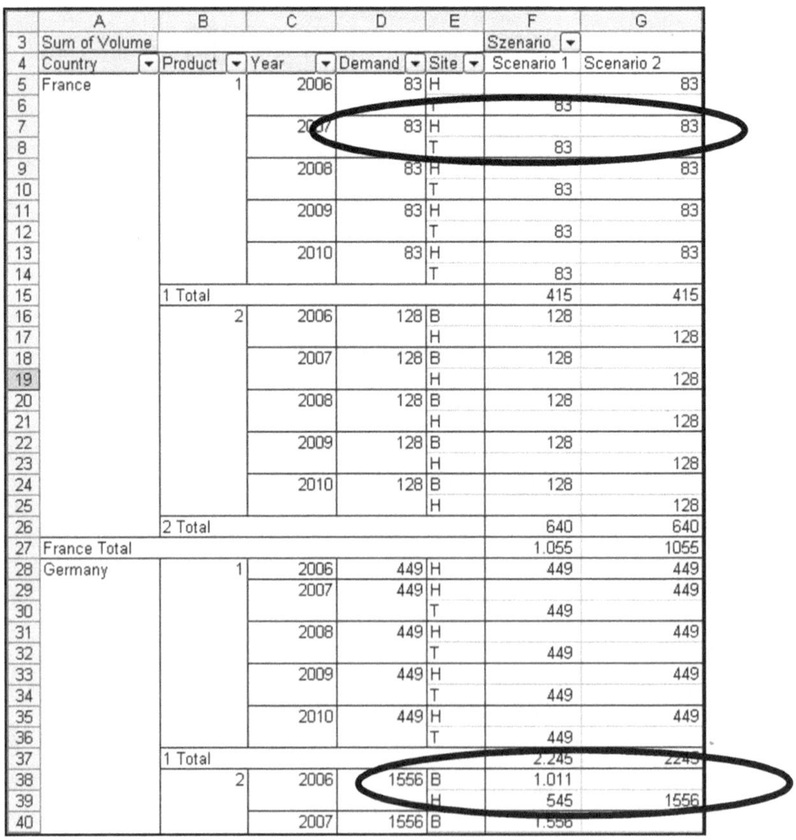

	A	B	C	D	E	F	G
3	Sum of Volume					Szenario	
4	Country	Product	Year	Demand	Site	Scenario 1	Scenario 2
5	France	1	2006	83	H		83
6					T	83	
7			2007	83	H		83
8					T	83	
9			2008	83	H		83
10					T	83	
11			2009	83	H		83
12					T	83	
13			2010	83	H		83
14					T	83	
15		1 Total				415	415
16		2	2006	128	B	128	
17					H		128
18			2007	128	B	128	
19					H		128
20			2008	128	B	128	
21					H		128
22			2009	128	B	128	
23					H		128
24			2010	128	B	128	
25					H		128
26		2 Total				640	640
27	France Total					1.055	1055
28	Germany	1	2006	449	H	449	449
29			2007	449	H		449
30					T	449	
31			2008	449	H		449
32					T	449	
33			2009	449	H		449
34					T	449	
35			2010	449	H		449
36					T	449	
37		1 Total				2.245	2245
38		2	2006	1556	B	1.011	
39					H	545	1556
40			2007	1556	B	1.556	

Fig. 52. Comparing results from alternative scenarios

5.5 Selected Findings from the Pilot Application

5.5.1 Reproducing the Status Quo to Obtain a Baseline

In order to obtain a baseline potential improvement measures can be compared to, the current network configuration was reproduced in a first step. To do so, the user only has to set single sourcing restrictions that reflect the current plant-market allocations for every product. Figure 53 shows the current product-plant allocations, the corresponding utilization of the installed technical capacity and the structure of the resulting operating cash flows for a pilot application at an aggregated level. The data was modified to maintain confidentiality but since alternative evaluations provided below were assessed in comparison to this baseline, the proportions of the changes are realistic.

Capacity utilization by product group (capacity units)

	Product group					Utilization
	1	2	3	4	5	
Site A	17.5	11.4		5.1	170.8	**85%**
Site B		9.1				**77%**
Site C		21.4				**86%**
Site D	5.8	2.8	1.4	0.6		**71%**
Site E		2.1		1.6		**79%**
Site F		3.2				**64%**

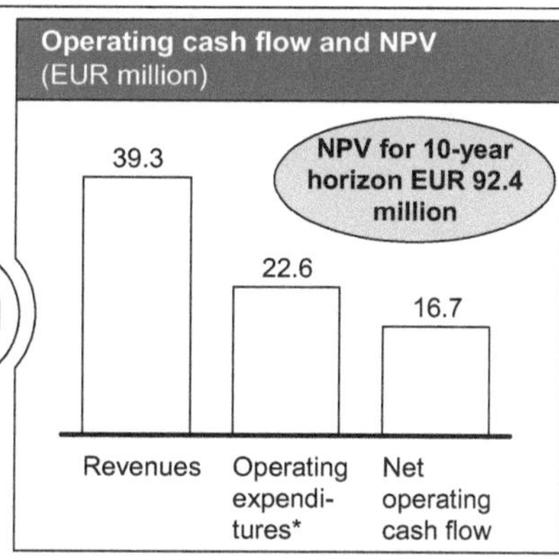

* Excluding restructuring and investment expenditures

Fig. 53. Baseline data

5.5.2 Assessing Alternative Environmental Scenarios

Using the Scenario Manager a broad range of analyses can be performed to assess the consequences of different market, cost or currency developments. To illustrate the insights that can be gained by performing these analyses, the example of an appreciation of the US$ compared to the Euro is chosen. As can be seen in the currency exposure chart in Figure 50, the value chain's business is dominated by home-currency denominated transactions (Euro). However, for both US$ and Chinese Remnibi a significant cost exposure exists. Due to the strong correlation between US$ and Remnibi, an appreciation of the US$ can considerably affect the profitability of the value chain. Figure 54 shows the comparison of the current exchange rate scenario with an appreciation of the US$ to the levels seen in 2001 relative to the Euro.

Without any changes to the production network, the operating cash flows and the NPV of the network would be reduced by approximately 10% in comparison to the baseline values. However, by re-allocating production volumes within existing capacities, it is possible to restore previously earned operating cash flows. To do so, production volumes are shifted to the major site A, which is located in the Euro zone. Contrarily, site C, which is located in the USA, would not be utilized at all by the product groups included in the example. It should be noted that this does not imply a closure of the US site since only a subset of the product portfolio was included in the analysis. The net present value of the network is nevertheless affected by the US$ appreciation because of the restructuring costs associated with the re-allocation of production volumes.

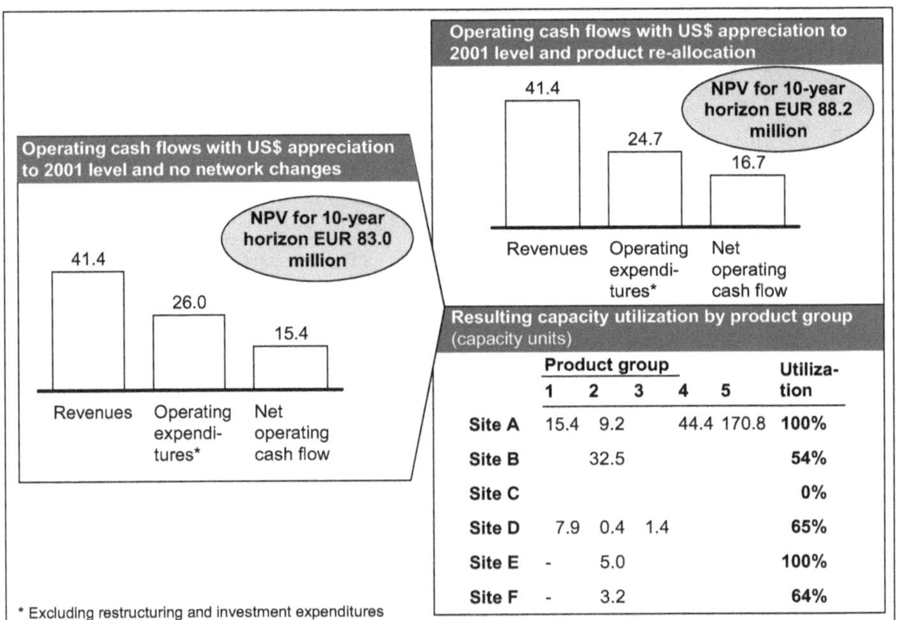

	Product group 1	2	3	4	5	Utilization
Site A	15.4	9.2		44.4	170.8	100%
Site B		32.5				54%
Site C						0%
Site D	7.9	0.4	1.4			65%
Site E	-	5.0				100%
Site F	-	3.2				64%

Fig. 54. Example of exchange rate effects analysis

5.5.3 Assessing Configuration Alternatives

While the optimization model is capable of assessing comprehensive "greenfield" production network redesigns, doing so requires the compilation of data on a large number of potential investment and restructuring alternatives. At the same time, implementing such a comprehensive redesign requires substantial capital expenditures and poses significant implementation challenges and risks. Hence, in practice it is often preferred to test the financial and operational effects of a limited number of discrete restructuring options that can be implemented step by step. As shown in Figure 53, with the exemplary subset of the product portfolio considered for the case study, site C (located in the US) only supplies a single product group. This product group is entirely sold in the US market. To illustrate the evaluation of discrete restructuring options, a configuration alternative to close the plant was created in order to assess the impact of a closure of the site. In addition to the overall financial impact the model also provides a recommendation on where to relocate the production volumes to. As can be seen in Figure 55, the US market would be supplied from three different sites simultaneously. The first of the three sites replacing the local production capacity is located in Europe while the other two sites are located in South America. Since in the baseline scenario existing capacities are not fully utilized, no new capacity has to be created and as the product group was already produced at the receiving sites no costs are incurred for the product transfer. The fixed costs savings from closing the US site by far offset the increases in transportation costs and import duties leading to a positive net present value of the restructuring option.

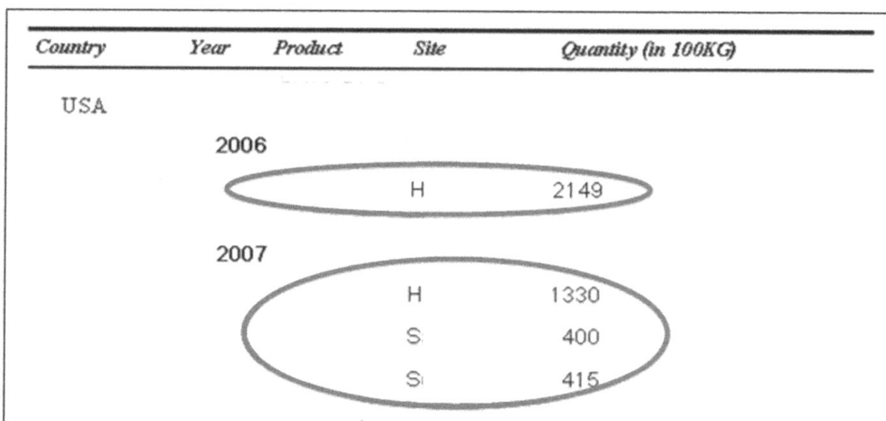

Fig. 55. Reallocation of production volumes after plant closure

6 Conclusion

Major players in specialty chemicals industry operate global production networks with numerous sites located in every economic area. If properly managed, these global production networks can by themselves be a source of competitive advantages e.g., via exploitation of factor cost advantages or providing operational hedging against currency fluctuations. However, in most cases today's production networks still lack the design required to do so. Among the historical reasons responsible for this gap are that:

- production capacities are concentrated within a few countries, primarily in Europe and North America, while economic growth in general and particularly demand growth for chemical products has shifted to the developing regions of Asia.
- the ongoing price-cost squeeze forces an industry that traditionally competed on product characteristics and was subsequently accustomed to high margins to focus on cost-efficient production.
- organic network growth often occurred in the form of market area plants that replicated production activities in new markets due to the significant trade restrictions that prevailed at the time capacity was added.
- the mergers and acquisitions leading to today's industry structure were often not accompanied by a comprehensive integration of the respective production networks.

To overcome these historical burdens companies have to re-design existing production networks and commence to employ an integrated approach towards managing their global production capacities. While managers in industry are aware of this necessity, due to the inherent complexity of the task and the lack of adequate decision support tools, comprehensive strategic network re-designs have so far been an exception. The objective of this work has been to develop a planning process and the analytical tools required to support industry in meeting this challenge. To ensure the practical relevance of the proposed processes and models, they were developed in close cooperation with a major specialty chemicals company that operates a production network containing more than 50 sites worldwide.

To lay the foundation, the role of production network design within supply chain management, the deficiencies of available commercial Advanced Planning Systems in the area of strategic supply network design, generic production network design principles and the relationship between production network design and industrial location science are discussed in Chapters 2.1 and 2.2. Additionally, peculiarities of specialty chemicals industry such as the diversity of the product portfolio and production processes and the interdependencies between the site and the value chain perspective are presented in Chapter 2.3. On this basis an integrated production network planning process is developed in Chapter 2.4 and its integration into overall strategic planning and controlling processes is discussed. The proposed production network planning process itself is split into two major phases: global value chain network optimization and site selection. This split accounts for the fact that the former is dominated by quantitative criteria while the latter is primarily driven by qualitative criteria. For each of these major phases a decision support tool tailored to the needs of specialty chemicals industry is developed.

With respect to the global value chain network optimization phase a review of the literature in Chapter 3.2 revealed that a majority of the models proposed so far rests on overly simplified assumptions and/or does not properly capture the economic aspects underlying production network design but instead focuses on more technical Operations Research aspects. Additionally, within the subset of application-oriented contributions none incorporating the specific needs of specialty chemicals industry was found. The Mixed-Integer Linear Programming model developed as a cornerstone of this work contributes towards closing this gap. It supports global production network design for individual specialty chemicals value chains. Based on a comprehensive discussion of alternative options to model the various elements of a production network in Chapter 3.3, formulations capturing the economic implications of alternative options, industry-specific requirements and the implications of global trade are proposed in Chapter 3.4. Major features of the model include:

- the approach to model technical capacity as a choice of alternative equipment size classes and degrees of automation,
- a separate modeling of capacity constraints for utilities shared by multiple plants operated at a site,
- an explicit consideration of the re-design of existing networks instead of solely modeling a green field approach,
- a modeling of the personnel capacity effects and the costs associated with personnel reductions caused by both changes to utilization and changes to technical capacity,

- the explicit consideration of the costs and capacity requirements associated with product transfers between sites and
- the incorporation of various aspects of international trade and investments such as different currencies, duties and duty drawbacks or investment subsidies.

Extensions to incorporate effects of taxation, an explicit modeling of product-mix effects (economies of scale and scope), make or buy/vendor selection and additional elements of technical capacity such as shared resources are provided in Chapter 3.4.3 to allow for an application to value chains with characteristics that differ from those that were considered when developing the model.

Furthermore, the implications of the end-of-horizon effect caused by using a net present value-based objective function in combination with a fixed planning horizon, mostly ignored in the academic literature, are discussed. To overcome this effect, a split of the planning horizon into two phases with investment decisions being limited to the first phase is suggested. This approach at the same time makes it possible to incorporate the effects of uncertainties into the model via so-called robust optimization techniques based on two-stage decisions with recourse. In contrast to stochastic optimization, which requires probability distributions for the uncertain parameters, the robust optimization approach presented can be integrated into the classical scenario planning approach commonly used in strategic planning.

Chapter 3.5 provides the results of numerical performance tests conducted which demonstrate that, despite its complexity, the resulting MILP model can be solved in reasonable time using the standard optimization software ILOG CPLEX 10.

The assessment of individual production sites, both in the context of site selection and site controlling, requires the simultaneous consideration of diverse, sometimes conflicting location factors that are to a large extent of qualitative nature. Hence, a Multiple Attribute Decision Analysis tool is most suitable to support this phase of the overall production network design process. Based on a review of the most common families of multiple attribute decision analysis methods and Data Envelopment Analysis, using the Analytic Hierarchy Process is suggested in Chapter 4. The decision support model proposed can be used both to support the site selection phase of a production network design project or as part of a regular site controlling process. Lessons learned from industry applications of the site assessment model illustrate its usefulness in practice and provide guidance on the relative importance of different location factors in specialty chemicals industry complementing the relatively old empirical findings.

Finally, Chapter 5 contains an application case study of the network design model describing how it was integrated into a decision support tool, providing step-by-step guidance on how to model a real-world production system and explaining the analyses the model can perform. The case study not only serves as a proof-of-concept for the model developed in this work but also intends to convey the value that employing Operations Research techniques can add in strategic planning processes. Due to the capability to perform scenario- and sensitivity analyses and assess the effects of both individual changes to a production network or structurally different design principles, the benefits of using such a model extend beyond the substantial economic improvements that can be achieved by a one-time optimization of historically grown production networks. Hopefully, the positive experiences gathered encourage decision makers in industry to use optimization models more frequently in production network design.

While the application case studies of the planning process and decision support models developed in this work demonstrate that significant improvements of specialty chemicals production networks can be achieved, several promising areas for further research remain. Among them the incorporation of uncertainty via the outlined robust optimization technique is probably the most prominent. Here, one aspect future research could focus on is how to derive alternative scenarios that capture the effects of uncertainty without knowing the underlying probability distributions. Another area requiring additional research is related to scale and scope economies associated with product-mix effects. While varyingly complex model formulations are available in literature, in practice it is so far not possible to quantify the required parameters with reasonable effort. Another area in need of future research is to integrate interdependencies between production network design and demand into network design models. In the model developed for this work it has been assumed that demand and prices are predetermined and independent of the cost position achieved by the company and the network design chosen. Market-related effects were instead captured via scenario analyses. However, in practice companies can frequently choose between market strategies associated with different success factors, prices and demand volumes. While the impact of network design on demand has been incorporated into distribution network design models for a long time (e.g., Tempelmeier 1980a; Tempelmeier 1980b) only two models incorporating this link into production network design were found. The model proposed by Chakravarty (2005) includes pricing options via price-demand functions but is only of limited value for practical applications. While Martel's approach (cf. Martel 2005) to include the choice between discrete marketing policy options is more application-oriented, it only takes into account the link between prices and delivery lead times.

Appendix

Appendix 1: Derivation of Discount Rate

As pointed out in Chapter 3.3.2 the company's cost of capital should be used to discount cash flows to their present value. As a company encounters different costs for equity and debt, the respective costs have to be identified as a prerequisite for calculating the cost of capital.

For publicly listed companies the Capital Asset Pricing Model (CAPM) is generally used to determine equity costs.[45] The CAPM calculates the minimum return investors require based on the non-diversifiable, systematic risk associated with an investment into the company's stock. This minimum return depends on the return on risk-free investments (e.g., government bonds), the market risk premium and the company-specific risk factor, which is usually referred to as beta factor. Beta describes a stock's sensitivity to market movements and is calculated by dividing the stock's covariance with the market by the variance of the market. The data required to calculate equity costs is provided by various financial services firms such as Ibbotson Associates, Bloomberg or Barra. The following formula calculates the cost of equity:

$$r = r_f + \beta(r_m - r_f) \tag{A.1}$$

where: r = Expected return on equity
r_f = Risk-free interest rate
r_m = Expected return of market portfolio
β = Company-specific risk factor

The costs of debt can generally be deduced from the terms of credit contracts or corporate bonds issued by the company. Alternatively, if a credit rating is available, the interest rate corresponding to the credit rating of the company can be used (cf. Koller et al. 2005, pp. 319-320).

---

[45] The paragraph is based on Brealey et al. (2006, pp. 189-204) and Koller et al. (2005, pp. 294-318).

The overall cost of capital are calculated using the Weighted Average Cost of Capital – Model (WACC) which weighs the cost of equity and cost of debt with the target capital structure (cf. Brealey et al. 2006, pp. 456-462, 503-520 and Koller et al. 2005, pp. 291-294). As interest paid on debt can be deducted from taxable income, pre-tax and after-tax cost of capital differ (the effect is also referred to as tax shield). Depending on whether pre-tax or after-tax cash flows are modeled, pre- or after-tax costs of capital have to be used correspondingly. Figure 56 provides an overview of the components of the WACC.

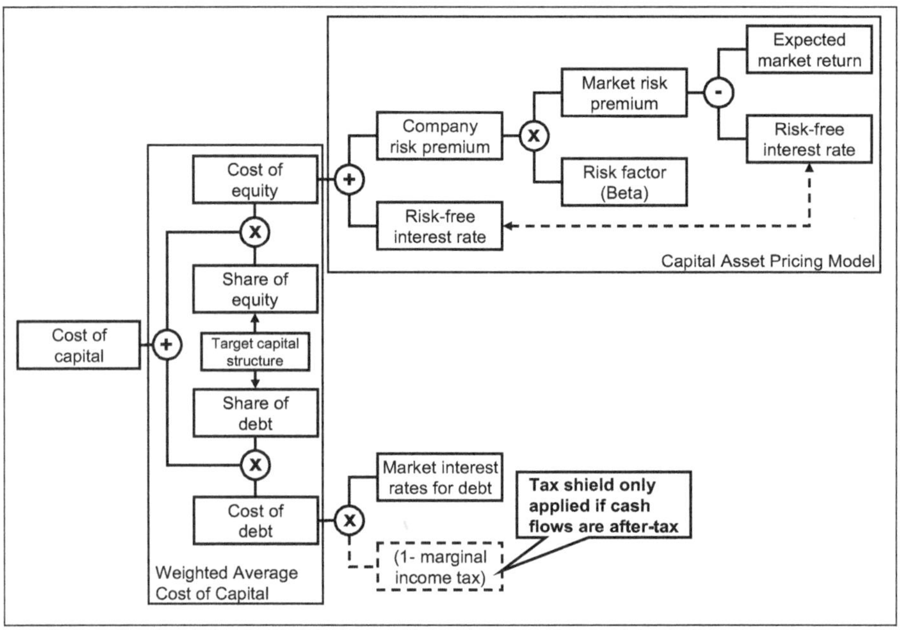

Fig. 56. Calculation of cost of capital[46]

The use of uniform costs of capital to evaluate all investment projects of a company implies that the risk associated with each individual investment equals the company's average risk. As this is often not the case in diversified companies, the use of more differentiated business unit or project specific cost of capital has been suggested in literature (cf. Brealey et al. 2006, pp. 215-218).

In the design of a global production network for a specific value chain additional risk stems from factors such as exchange rate exposure or political risks associated with international investments. Following the differen-

[46] Based on Hahn and Hungenberg (2001), p. 160.

tiated cost of capital logic, adjusting the discount rate has been suggested as an appropriate method to incorporate these risks into the analysis by some authors (e.g., Micallef 1981, p. 50), differentiated discount rates are commonly used in practice (cf. Hayes and Wheelwright 1984, p. 144) and hence some production network design models also employ location-specific discount rates (cf. Haug 1985, p. 85).

However, the risks to be considered usually are not of a systematic nature (for a discussion of risks that affect systematic risk see Butler and Joaquin 1998). To the contrary, international diversification may even contribute to reducing a company's systematic risk (cf. Agmon and Lessard 1977). Furthermore, determining appropriate discount rate adjustments is difficult and frequently leads to an overestimation of the effects of risk because of the large impact minor increases of the discount rate have on the NPV (cf. Brewer 1981, p. 10). Accordingly, authors such as Feils and Sabac (2000, p. 133), Brewer (1981, p. 10), Shapiro (1981, p. 64) and Shapiro (1978, p. 9) strongly argue that a uniform discount rate should be used for multinational operations and risks associated with specific investments should be accounted for by adjusting expected cash flows.

Appendix 2: Tariff Regulations[47]

Tariffs are import taxes levied at the border. They generally have two purposes: protecting local industry and providing revenues for the government. Especially for less developed countries, tariffs are an important source of tax income. Tariffs can take different forms. *Specific tariffs* are flat rates on particular products (e.g., 200 EUR per ton). Variable tariffs take account of changes between domestic and world market prices (e.g., agricultural tariffs in the European Union). The most common form of tariffs is to calculate a percentage of the price of the imported product (*ad valorem*).

Tariff rates are set on a product/product category level. The classification of products is in many countries based on a harmonized customs nomenclature and a nominal tariff is set for each product category. Tariff structures however often contain temporary or permanent exceptions, preferential rates for less developed countries, etc. Therefore, one has to distinguish between the nominal tariff and the tariff actually applied.

An important aspect of ad valorem tariff regimes is the procedure for determining the value of a product. In the past, countries employed differ-

[47] This appendix is based on Fraedrich (2001); McDonald (1997) and Jackson (1997).

ent valuation rules. For example, the United States valued imported chemical products with the American selling price. Valuation rules were harmonized with the Tokyo and Uruguay rounds of the GATT negotiations (General Agreement on Tariffs and Trade). The value is generally determined based on the transaction value. Only if this is not possible can the other five methods be applied in the prescribed hierarchical order:

1. The transaction value of the imported goods
2. Transaction value of identical goods
3. Transaction value of similar goods
4. Deductive method (adjusted selling price in country of import)
5. Computed method (value calculated from production costs)
6. Fall-back-method

The large majority of tariffs are in the range of 0 to 10% of the product's value. However, in extreme cases, tariffs of more than 30% might occur. Many tariff structures contain an escalation process whereby tariffs on raw materials are lower than those on goods further processed from that raw material. In many countries basic raw materials have no tariff at all. Also, if products are re-exported either without transformation or after transformation, many governments offer tariff drawbacks effectively paying back the tariffs levied on any raw materials or components imported to produce the product.

Between the members of the GATT agreement a Most Favored Nation (MFN) policy generally prohibits the application of tariffs that discriminate against certain countries. However, as there are exceptions possible to the MFN policy (e.g., for free trade zones) and some countries are not members of GATT, *rules of origin* are an important determinate of the actual tariff rate. Unfortunately, there is no uniform definition of what "origin" means but each country has the right to define its own rules of origin.

Rules of origin applied within free trade zones in many cases differ from those generally applied. Also, preferential tariffs for least-developed countries are commonly linked to stricter rules of origin. Therefore, one has to distinguish between *preferential* rules of origin (where exceptions apply) and *non-preferential* rules of origin. For non-preferential rules of origin a harmonization effort is ongoing at the World Trade Organization (WTO). Preferential rules of origin can take specific forms for each underlying trade agreement. Their definitions are most critical when the objective is to exploit free trade zones such as NAFTA by locating certain production tasks in one of the member countries while still producing a significant amount of intermediates/components in a country that is not a member of the free trade zone.

It is common to distinguish between three types of products. Goods wholly the produce of a country originate only from that country (e.g., unmanufactured raw products). The other two groups are goods wholly manufactured in the country from specific materials and goods partly manufactured in the country. The determination of origin is only critical for the last group.

The *substantial transformation* principle states that a product is attributed to the most recent exporting country only if within that country there has been a substantial transformation of the input goods obtained from another country. Several principles are widely used to determine the origin of goods:

- One indicator is that the transformation leads to a *change in tariff classification*. This sometimes leads to arbitrary results because different parts of the tariff nomenclature have different levels of detail.
- Based on the *value-added* approach goods are attributed to the last country of export if that country has added a minimum percentage of value to the goods. *Local content rules* belong to the value-added approach. They focus on the value of materials included in a product and set a minimum requirement for materials originating from the respective country.
- For chemical products a *process rule* is applied, too. It states that a product is originating from a country if the chemical reaction creating the product occurred in that country.

Exceptions from the substantial transformation principle can be found for example in the form of outward processing rules or integrated sourcing initiatives. If the importer is the country's government, *offset requirements* can be established to limit the outflow of cash. These requirements force the exporter to purchase a certain value of products produced in the importing country, to establish local manufacturing or to use locally produced components in manufacturing. This type of restriction is especially common in the context of military purchases and not relevant for specialty chemicals production.

Appendix 3: Political Risk

Political risks have to be considered both in the case of identifying potential plant locations and in the context of production network controlling. Both in academic research and in industry the Iranian revolution jump-started comprehensive discussions of the subject (cf. Brewer 1981, p. 5;

Micallef 1981, p. 47; Chase et al. 1988, p. 35; Weiner 1992, p. 20). Nevertheless, according to Kobrin (1979, p. 75) political risks are rarely included explicitly into the analysis of major investment projects.

Different approaches towards structuring the various kinds of political risks can be found in literature (e.g., Meldrum 2000, pp. 34-35; Desta 1986, pp. 50-53). As shown in Figure 57, typically, based on the impact of the respective type of political risk, one can distinguish between asset protection risks, transfer risks and operational risks (cf. Root 1972, p. 357). Additionally, a separation into macro risks affecting all foreign companies operating in a country and micro risks that are specific to the industry and/or the individual company should be made (cf. Robock 1971, pp. 9-11).

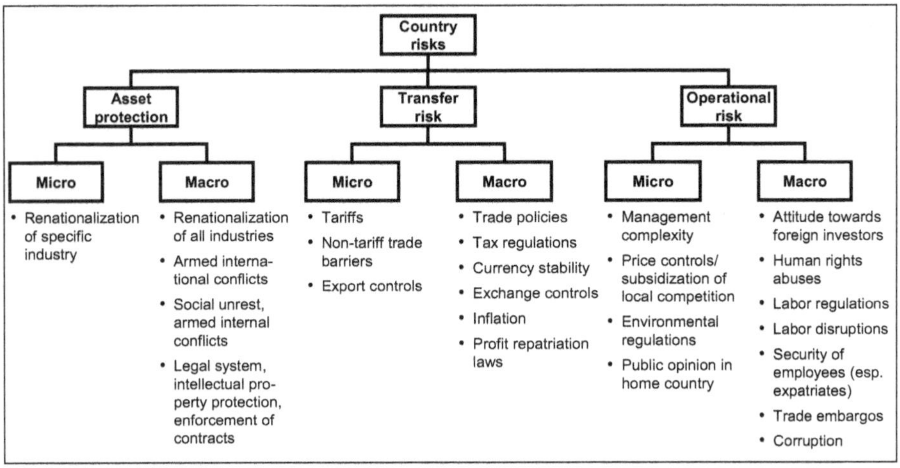

Fig. 57. Overview country-specific risks

The relative importance of the different risk components depends on the company or the individual investment project. In literature, because of the dramatic effects, typically macro risks and especially expropriation risks have received considerable attention (cf. Shapiro 1981, p. 63; Brewer 1981, p. 5; Hofer and Haller 1980, p. 43). The latter has however been of diminishing importance during recent years (cf. Minor 1994) even though expropriations of foreign investors recently occurred both in Russia (cf. Feils and Sabac 2000, pp. 139-141; Wilkin 2000, pp. 42-43) and in the oil industry in Latin America. Nevertheless, in the context of operating global specialty chemicals production networks a comprehensive analysis of operational risks appears to be more important.

In a first step the criteria to be considered have to be established. Subsequently, appropriate data sources have to be identified. If internal data sources are to be used, according to Hofer and Haller (1980, pp. 44-45), a common approach is to rely on:

- *Grand Tours*, where decision makers visit all countries under consideration.
- *Old Hands*, where persons with substantial experiences in the relevant countries are interviewed.
- *Delphi-method*, where an expert panel conducts a systematic analysis of the different countries.

Additionally, various research companies provide rankings and comprehensive country analyses. However, these often cover only a small subsegment of the relevant micro risks and the components of the overall score and the respective component weights are often not revealed (cf. Desta 1986, pp. 45-47; Meldrum 2000, pp. 33-34). Krystek and Walldorf (1997) discuss seven commonly used country ratings. The *Index of Economic Freedom*, published annually by the *Heritage Foundation* and *The Wall Street Journal*, provides data on aspects such as trade policy, corruption, legal system and labor market regulations. In contrast to other sources, the electronic version can be downloaded free of charge (www.heritage.org) and the detailed criteria used to assess the more than 160 countries covered are provided (cf. Beach and Miles 2006). Focusing on the small-scale, high frequency risks originating from a lack of transparent and generally accepted principles governing the relationships between businesses, investors and governments, Kurtzman et al. (2004) publish *The Opacity Index*. The 2004 version of this index can also be downloaded free of charge at www.opacity-index.com. Another index that is used frequently, the *BERI-Index* (Business Environment Risk Information) is provided by BERI S.A. Hake (1982) discusses its components but it has to be assumed that many changes occurred in the meantime.

References

Abdallah WM (1989) International transfer pricing policies. Quorum Books, Westport and London

Adler N, Friedman L, Sinuany-Stern Z (2002) Review of ranking methods in the data envelopment analysis context. European Journal of Operational Research 140: 249-265

Agmon T, Lessard DR (1977) Investor recognition of corporate international diversification. The Journal of Finance 32: 1049-1055

Ahmed S, Sahinidis NV (1998) Robust Process Planning under Uncertainty. Industrial Engineering & Chemistry Research 37: 1883-1892

Aikens CH (1985) Facility location models for distribution planning. European Journal of Operational Research 22: 263-279

Ambrosino D, Scutellà MG (2005) Distribution network design: New problems and related models. European Journal of Operational Research 165: 610-624

Amecke H-B (1987) Chemiewirtschaft im Überblick. VCH, Weinheim et al.

Andersen P, Petersen NC (1993) A Procedure for Ranking Efficient Units in Data Envelopment Analysis. Management Science 39: 1261-1264

Anderson SP, Schmitt N (2000) Non-tariff barriers and trade liberalization. ifo Institut, München

Anderson SW (1995) Measuring the Impact of Product Mix Heterogeneity on Manufacturing Overhead Cost. The Accounting Review 70: 362-387

Anthony RN (1965) Planning and Control Systems: A Framework for Analysis. Harvard University, Boston

Antunes A, Peeters D (2001) On solving complex multi-period location models using simulated annealing. European Journal of Operational Research 130: 190-201

Arntzen BC, Brown GG, Harrison TP, Trafton LL (1995) Global Supply Chain Management at Digital Equipment Corporation. Interfaces 25: 69-93

Artikis GP (1991) Plant Location Decisions in the Greek Food Industry. International Journal of Operations & Production Management 11: 57-70

Autschbach J (1997) Internationale Standortwahl. Deutscher Universitäts-Verlag, Wiesbaden

Ballou RH (1992) Business Logistics Management. Prentice-Hall, Englewood Cliffs

Ballou RH, Masters JM (1999) Facility Location Commercial Software Survey. Journal of Business Logistics 20: 215-232

Bana e Costa CA, De Corte J-M, Vansnick J-C (2005) On the mathematical foundations of MACBETH. In: Figueira J, Greco S, Ehrgott M (eds) Multiple Criteria Decision Analysis: State of the Art Surveys. Springer, Berlin et al., pp 409-442

Bana e Costa CA, Ensslin L, Corrêa EC, Vansnick JC (1999) Decision Support Systems in action: Integrated application in a multicriteria decision aid process. European Journal of Operational Research 113: 315-335

Bankhofer U (2003) Standortprobleme im Industrieunternehmen - Entwicklung und Ableitung geeigneter Planungsmodelle und -methoden auf Basis eines Kategorisierungsansatzes. Zeitschrift für Planung 14: 25-50

Bankhofer U (2001) Industrielles Standortmanagement. Deutscher Universitäts-Verlag, Wiesbaden

Bankhofer U (2000) Industrielles Standortmanagement - Ergebnisse einer empirischen Untersuchung. Zeitschrift für Planung 11: 329-352

Bankhofer U Kugler M (1999) Eine empirische Analyse zum Standortmanagement industrieller Großunternehmen. Institut für Statistik und Mathematische Wirtschaftstheorie Universität Augsburg, Augsburg

Bartels E, Claret J, Deppe S, Marquardt R (2006) Survival when It's Hard to be Special: A Perspective on Specialty Chemicals. In: Budde F, Felcht U-H, Frankemölle H (eds) Value Creation: Strategies for the Chemical Industry, 2nd edn. Wiley-VCH, Weinheim, pp 95-108

Bartmess A, Cerny K (1993) Building Competitive Advantage through a Global Network of Capabilities. California Management Review 35: 8-103

Bass BM, McGregor DW, Walters JL (1977) Selecting Foreign Plant Sites: Economic, Social and Political Considerations. Adacemy of Management Journal 20: 535-551

Bauer A, Domscke W (1993) Competitive location on a network. European Journal of Operational Research 66: 372-391

Bea FX (2000) Entscheidungen des Unternehmens. In: Bea FX, Dichtl E, Schweitzer M (eds) Allgemeine Betriebswirtschaftslehre, 8th edn. Lucius & Lucius, Stuttgart, pp 303-410

Beach WW, Miles MA (2006) Explaining the Factors of the Index of Economic Freedom. In: Miles MA, Feulner EJ, O'Grady MA (eds) 2006 Index of Economic Freedom. The Heritage Foundation & Dow Jones & Company, Inc., Washington, D.C., pp 55-76

Beamon BM (1998) Supply chain design and analysis: Models and methods. International Journal of Production Economics 55: 281-294

Beckman SL (1990) Manufacturing flexibility: The next source of competitive advantage. In: Moody PE (ed) Strategic Manufacturing. Dow-Jones Irwin, Homewood, pp 107-132

Beckman SL, Boller WA, Hamilton SA, Monroe JW (1990) Using manufacturing as a competitive weapon: The development of a manufacturing strategy. In: Moody PE (ed) Strategic Manufacturing. Dow-Jones Irwin, Homewood, Illinois, pp 53-75

Behrens KC (1971) Allgemeine Standortbestimmungslehre, 2nd edn. Westdeutscher Verlag, Opladen

Behrens KC (1960) Zur Typologie und Systematik der Standortlehren. In: Stammer O, Thalheim KC (eds) Festgabe für Friedrich Bülow. Duncker & Humblot, Berlin, pp 51-61

Belton V (1986) A comparison of the analytic hierarchy process and a simple multi-attribute value function. European Journal of Operational Research 26: 7-21

Belton V, Hodgkin J (1999) Facilitators, decision makers, D.I.Y. users: Is intelligent multicriteria decision support for all feasible or desirable? European Journal of Operational Research 113: 247-260

Belton V, Stewart TJ (2002) Multiple Criteria Decision Analysis. Kluwer Academic Publishers, Boston et al.

Belton V, Vickers SP (1993) Demystifying DEA - A Visual Interactive Appproach Based on Multiple Criteria Analysis. The Journal of the Operational Research Society 44: 883-896

Belton V, Vickers SP (1989) V I S A - VIM for MCDA. In: Locket AG, Islei G (eds) Improving Decision Making in Organizations. Springer, Berlin et al., pp 287-304

Benjaafar S, Gupta D (1999) Workload allocation in multi-product, multi-facility production systems with setup times. IIE Transactions 31: 339-352

Berry WL, Hill TJ, Klompmaker JE (1995) Customer-driven manufacturing. International Journal of Operations & Production Management 15: 4-15

Bertrand JWM (2003) Supply Chain Design: Flexibility Considerations. In: de Kok AG, Graves SC (eds) Supply Chain Management: Design, Coordination and Operation. Elsevier, Amsterdam et al., pp 133-198

Bestmann U (2001): Der Gegenstand der Betriebswirtschaftslehre, Betriebswirtschaftlicher Gesamtprozeß und Konstitutiver Rahmen. In: Bestmann U (ed) Kompendium der Betriebswirtschaftslehre, 10th edn. R. Oldenburg, München, Wien, pp 1-74

Bhutta KS (2004) International facility location decisions: a review of the modelling literature. International Journal of Integrated Supply Chain Management 1: 33-50

Bhutta KS, Huq F, Frazier G, Mahamed Z (2003) An integrated location, production, distribution and investment model for a multinational corporation. International Journal of Production Economics 86: 201-216

Bienstock D, Shapiro JF (1988) Optimizing resource acquisition decisions by stochastic programming. Management Science 34: 215-229

Billington CA, Davis TC (1992) Manufacturing Strategy Analysis: Models and Practice. Omega International Journal of Management Science 20: 587-595

Birge JR, Louveaux F (1997) Introduction to Stochastic Programming. Springer, Berlin et al.

Bitran GR, Gilbert SM (1990) Sequencing Production on Parallel Machines with Two Magnitudes of Sequence-dependent Setup Cost. Journal of Manufacturing and Operations Management 3: 24-52

Bitran GR, Tirupati D (1993) Hierarchical Production Planning. In: Graves SC, Rinnooy Kan AHG, Zipkin PH (eds) Logistics of Production and Inventory. Elsevier Science Publishers, Amsterdam, pp 523-568

Blair JP, Premus R (1987) Major Factors in Industrial Location: A Review. Economic Development Quarterly 1: 72-85

Blömer F (1999) Produktionsplanung und -steuerung in der chemischen Industrie. Gabler, Wiesbaden

Bok J-K, Lee H, Park S (1998) Robust investment model for long-range capacity expansion of chemical processing networks under uncertain demand forecast scenarios. Computers & Chemcial Engineering 22: 1037-1049

Borcherding K, Eppel T, von Winterfeldt D (1991) Comparison of Weighting Judgements in Multiattribute Utility Measurement. Management Science 37: 1603-1619

Bottomley PA, Doyle JR (2001) A comparison of three weight elicitation methods: good, better, and best. Omega International Journal of Management Science 29: 553-560

Bouyssou D (1999) Using DEA as a tool for MCDM: some remarks. Journal of the Operational Research Society 50: 974-978

Bramel J, Simchi-Levi D (1997) The Logic of Logistics. Springer, Berlin et al.

Brandeau ML, Chiu SS (1989) An Overview Of Representative Problems In Location Research. Management Science 35: 645-674

Brans JP, Mareschal B (2005) Promethee method. In: Figueira J, Greco S Ehrgott M (eds) Multiple Criteria Decision Analysis: State of the Art Surveys. Springer, Berlin et al., pp 163-195

Brans JP, Mareschal B (1990) The PROMETHEE methods for MCDM; the PROMCALC, GAIA and Bankadvisor software. In: Bana e Costa CA (ed): Readings in Multiple Criteria Decision Aid. Springer, Berlin et al., pp 216-252

Brans JP, Mareschal B, Vincke P (1984) PROMETHEE: A new family of outranking methods in multicriteria analysis. Operational Research '84, pp 477-490

Brans JP, Vincke P (1985) A Preference Ranking Organisation Method. Management Science 31: 647-656

Brans JP, Vincke P, Mareschal B (1986) How to select and how to rank projects: The PROMETHEE method. European Journal of Operational Research 24: 228-546

Brealey RA, Myers SC, Allen F (2006) Corporate Finance, 8th edn. McGraw Hill, Boston et al.

Brede H. (1971) Bestimmungsfaktoren industrieller Standorte. John Wiley & Sons, Berlin

Breitman RL, Lucas JM (1987) PLANETS: A Modeling System for Business Planning. Interfaces 17: 94-106

Brewer TL (1981) Political Risk Assessment for Foreign Direct Investment Decisions: Better Methods for Better Results. Columbia Journal of World Business 16: 5-12

Brown GG, Graves GW, Honczarenko MD (1987) Design and operation of a multicommodity production/distribution system using primal goal decomposition. Management Science 33: 1469-1480

Brown RW, Northup WD, Shapiro JF (1986) LOGS: A modeling and optimization system for business planning. In: Mitra G (ed) Computer Assisted Decision Making. North-Holland Publishing Company, Amsterdam et al., pp 227-241

Brownlow SA, Watson SR (1987) Structuring Multi-Attribute Value Hierarchies. The Journal of the Operational Research Society 38: 309-317

Brugha CM (1998) Structuring and Weighting Criteria in Multi Criteria Decision Making (MCDM). In: Stewart TJ, van den Honert RC (eds) Trends in Multicriteria Decision Making. Springer, Berlin et al., pp 229-242

Brush TH, Maritan CA, Karnani A (1999) The plant location decision in multinational manufacturing firms: an empirical analysis of international business and manufacturing strategy perspectives. Production and Operations Management 8: 109-132

Bürklin W (1993) Die vier kleinen Tiger. Ullstein, Frankfurt/Main and Berlin

Butler KC, Joaquin DC (1998) A Note on Political Risk and the Required Return on Foreign Direct Investment. Journal of International Business Studies 29: 599-607

Camm JD, Chorman TE, Dill FA, Evans JR, Sweeney DJ, Wegryn GW (1997) Blending OR/MS, Judgment, and GIS: Restructuring P&G's Supply Chain. Interfaces 27: 128-142

Canel C, Das SR (2002) Modeling global facility location decisions: integrating marketing and manufacturing decisions. Industrial Management & Data Systems 102: 110-118

Canel C, Das SR (1999) The uncapacitated multi-period facilities location with profit maximization. International Journal of Physical Distribution & Logistics 29: 409-433

Canel C, Khumawala BM (2001) International facilities location: a heuristic procedure for the dynamic uncapacitated problem. International Journal of Production Research 39: 3975-4000

Canel C, Khumawala BM (1997) Multi-period international facilities location: an algorithm and application. International Journal of Production Research 35: 1891-1910

Canel C, Khumawala BM (1996) A mixed-integer programming approach for the international facilities location problem. International Journal of Operations & Production Management 16: 49-68

Canel C, Khumawala BM, Law J, Loh A (2001) An algorithm for the capacitated, multi-commodity multi-period facility location problem. Computers & Operations Research 28: 411-427

Chakravarty AK (2005) Global plant capacity and product allocation with pricing decisions. European Journal of Operational Research 165: 157-181

Chakravarty AK (1999) Profit margin, process improvement and capacity decisions in global manufacturing. International Journal of Production Research 37: 4235-4257

Chambers S (1992) Flexibility in the context of manufacturing strategy. In: Voss CA (ed) Manufacturing Strategy. Chapman & Hall, New York et al., pp 283-295

Chardaire P, Sutter A, Costa M-C (1996) Solving the Dynamic Facility Location Problem. Networks 28: 117-124

Charnes A, Cooper WW, Lewin AY, Seiford LM (1994) Data Envelopment Analysis: Theory, Methodoloy, and Application. Kluwer Academic Publishers, Boston et al.

Charnes A, Cooper WW, Rhodes E (1978) Measuring the efficiency of decision making units. European Journal of Operational Research 2: 429-444

Chase CD, Kuhle JL, Walther CH (1988) The Relevance of Political Risk in Direct Foreign Investment. Management International Review 28: 31-38

Chase RB, Jacobs RF, Aquilano NJ (2006) Operations management for competitive advantage, 11th edn. McGraw Hill, Boston et al.

Chen F (2003) Information Sharing and Supply Chain Coordination. In: de Kok AG, Graves SC (eds) Supply Chain Management: Design, Coordination and Operation. Elsevier, Amsterdam et al., pp 341-421

Cheng EWL, Li H, Yu L (2005) The analytic network process (ANP) approach to location selection: a shopping mall illustration. Construction Innovation 5: 83-97

Chernotsky HI (1983) Selecting U.S. Sites: A Case Study of German and Japanese Firms. Management International Review 23: 45-55

Choi FDS, Frost CA, Meek GK (2002) International Accounting. Prentice-Hall, Upper Saddle River

Chopra S, Meindl P (2004) Supply Chain Management, 2nd edn. Prentice-Hall, Englewood Cliffs

Christopher M (2005) Logistics and Supply Chain Management, 3rd edn. Financial Times Prentice Hall, Harlow et al.

Cohen MA, Fisher M, Jaikumar R (1989) International manufacturing and distribution networks: a normative model framework. In: Ferdows K, (ed) Managing international manufacturing. Elsevier Science Publishers, Amsterdam, pp 67-93

Cohen MA, Huchzermeier A (1999) Global supply chain management: a survey of research and applications. In: Tayur S, Ganeshan R, Magazine M (eds) Quantitative models for supply chain management. Kluwer Academic Publishers, Boston et al., pp 669-702

Cohen MA, Kleindorfer PR (1993) Creating value through operations: the lecacy of Elwood S. Buffa. In: Sarin RK (ed) Perspectives in operations management: Essays in honor of Elwood S. Buffa. Kluwer Academic Publishers, Boston et al., pp 3-21

Cohen MA, Lee HL (1989) Resource Deployment Analysis of Global Manufacturing and Distribution Networks. Journal of Manufacturing and Operations Management 2: 81-104

Cohen MA, Lee HL (1985) Manufacturing strategy. In: Kleindorfer PR (ed) The Management of Productivity and Technology in Manufacturing. Plenum Press, New York and London, pp 153-188

Cohen MA, Mallik S (1997) Global supply chains: research and applications. Production and Operations Management 6: 193-210

Cohen MA, Moon S (1991) An integrated plant loading model with economies of scale and scope. European Journal of Operational Research 50: 266-279

Cohen MA, Moon S (1990) Impact of Production Scale Economies, Manufacturing Complexity, and Transportation Costs on Supply Chain Facility Networks. Journal of Manufacturing and Operations Management 3: 269-292

Cokins G (2001) Activity-based cost management. John Wiley & Sons, New York et al.

Cooper R, Kaplan RS (1998) The Design of Cost Management Systems, 2nd edn. Prentice-Hall, Upper Saddle River

Cooper WW (2001) Data Envelopment Analysis. In: Gass SI, Harris CM (eds) Encyclopedia of Operations Reserach and Management Science, 2nd edn. Kluwer Academic Publishers, Boston et al., pp 183-191

Copeland TE, Joshi Y (1996) Why derivatives don't reduce FX risk. The McKinsey Quarterly 1996: 66-79

Corsten H, May C (1994) Besonderheiten der Produktion in der Verfahrensindustrie und ihre Auswirkungen auf PPS-Systeme. In: Corsten, H. (ed) Handbuch Produktionsmanagement. Gabler, Wiesbaden, pp 871-889

Current J, Min H, Schilling D (1990) Multiobjective analysis of facility location decisions. European Journal of Operational Research 49: 295-307

Dangayach GS, Deshmukh SG (2001) Manufacturing strategy: Literature review and some issues. International Journal of Operations & Production Management 21: 884-932

Dantzig GB (1955) Linear Programming Under Uncertainty. Management Science 1: 197-206

Dasci A, Verter V (2001) The plant location and technology acquisition problem. IIE Transactions 33: 963-973

Daskin MS, Snyder LV, Berger RT (2005) Facility location in supply chain design. In: Langevin A, Riopel D (eds) Logistics systems: Design and optimization. Springer, Berlin et al., pp 39-65

Dasu S, de la Torre J (1997) Optimizing an International Network of Partially Owned Plants Under Conditions of Trade Liberalization. Management Science 43: 313-333

Dasu S, Li L (1997) Optimal Operating Policies in the Presence of Exchange Rate Variability. Management Science 43: 705-722

De Meyer A, Vereecke A (2001) International operations. In: Warner M (ed) International Encyclopedia of Business and Management, 2nd edn. Thomson, London et al., pp 3349-3364

Desta A (1986) Assessing political risk in less developed countries. Journal of Business Strategy 5: 40-53

Dixit AK, Pindyck RS (1995) The Options Approach to Capital Investment. Harvard Business Review 73: 105-115

Dodgson J, Spackman M, Pearman A, Phillips L (2000) DTLR multi-criteria analysis manual. Department for Transport, Local Government and the Regions, London

Dogan K, Goetschalckx M (1999) A primal decomposition method for the integrated design of multi-period production-distribution systems. IIE Transactions 31: 1027-1036

Domschke W, Drexl A (1985) Location and Layout Planning. Springer, Berlin et al.

Domschke W, Krispin G (1997) Location and layout planning. OR Spectrum 19: 181-194

Dormer A, Vazacopoulos A, Verma N, Tipi H (2005) Modeling & solving stochastic programming problems in supply chain management using Xpress-SP. In: Geunes J, Pardalos PM (eds) Supply Chain Optimization. Springer, Berlin et al., pp 307-354

Dornier P-P, Ernst R, Fender M, Kouvelis P (1998) Global Operations and Logistics. John Wiley & Sons, New York et al.

Doyle J, Green R (1993) Data Envelopment Analysis and Multiple Criteria Decision Making. Omega International Journal of Management Science 21: 713-715

Drexl A, Fleischmann B, Günther H-O, Stadtler H, Tempelmeier H (1994) Konzeptionelle Grundlagen kapazitätsorientierter PPS-Systeme. Zeitschrift für Betriebswirtschaftliche Forschung 46: 1022-1045

Drezner Z (1995) Facility Location. Springer, Berlin et al.

Drezner Z, Hamacher HW (2001) Facility Location. Springer, Berlin et al.

Drezner Z, Klamroth K, Schöbel A, Wesolowsky GO (2001) The Weber Problem. In: Drezner Z, Hamacher HW (eds) Facility Location. Springer, Berlin et al., pp 1-36

Dufey G, Mirus R (1981) Forecasting Foreign Exchange Rates: A Pedagogical Note. Columbia Journal of World Business 16: 53-61

Dyer JS (1990) Remarks on the Analytic Hierarchy Process. Management Science 36: 249-258

Eckenrode RT (1965) Weighting multiple criteria. Management Science 12: 180-192

Ehrfeld W, Hessel V, Löwe H (2000) Microreactors. Wiley-VCH, Weinheim

Eiselt HA (1992) Location modeling in practice. American Journal of Mathematical and Management Sciences 12: 3-18

Eiselt HA, Laporte G (1995) Objectives in Location Problems. In: Drezner Z (ed) Facility Location. Springer, Berlin et al., pp 151-180

Eiteman DK, Stonehill AI, Moffett MH (2006) Multinational Business Finance, 11th edn. Pearson Education, Boston et al.

Emmrich V (1997) Globale Produktionsstandortstrategien. In: Krystek U, Zur E (eds) Internationalisierung. Springer, Berlin et al., pp 233-248

Eppen GD, Martin RK, Schrage L (1989) A scenario approach to capacity planning. Operations Research 37: 517-527

Erkut E, Neuman S (1989) Analytical models for locating undesirable facilities. European Journal of Operational Research 40: 275-291

Erlenkotter D (1981) A comparative study of approaches to dynamic location problems. European Journal of Operational Research 6: 133-143

Ernst & Young (2003) Transfer Pricing 2003 Global Survey. Ernst & Young

Fandel G, Stammen M (2004) A general model for extended strategic supply chain management with emphasis on product life cycles including development and recycling. International Journal of Production Economics 89: 293-308

Feils DJ, Sabac FM (2000) The impact of political risk on the foreign direct investment decision: A capital budgeting analysis. The Engineering Economist 45: 129-143

Feinschreiber R (2004) Transfer pricing methods. John Wiley & Sons, Hoboken, N.J.

Feinschreiber R (2000) Transfer pricing international. John Wiley & Sons, New York et al.

Feinschreiber R, Kent M (2003) Transfer pricing for intercompany transactions. In: Choi FDS (ed) International Finance and Accounting Handbook, 3rd edn. John Wiley & Sons, New York et al. pp 29.1-29.40

Ferdows K (1997a) Made in the world: the global spread of production. Production and Operations Management 6: 102-109

Ferdows K (1997b) Making the Most of Foreign Factories. Harvard Business Review 75: 73-88

Ferdows K (1989) Mapping international factory networks. In: Ferdows, K (ed) Managing international manufacturing. Elsevier Science Publishers, Amsterdam, pp 3-21

Figueira J, Greco S, Ehrgott M (2005a) Introduction. In: Figueira J, Greco S, Ehrgott M (eds) Multiple Criteria Decision Analysis: State of the Art Surveys. Springer, Berlin et al., pp XXI-XXXVI

Figueira J, Mousseau V, Roy B (2005b) ELECTRE methods. In: Figueira J, Greco S, Ehrgott M (eds) Multiple Criteria Decision Analysis: State of the Art Surveys. Springer, Berlin et al., pp 133-162

Flaherty TM (1986) Coordinating International Manufacturing and Technology. In: Porter ME (ed) Competition in Global Industries. Harvard Business School Press, Boston, MA, pp 83-109

Fleischmann B, Ferber S, Henrich P (2006) Strategic Planning of BMW's Global Production Network. Interfaces 36: 194-208

Fleischmann B, Meyr H (2003) Planning Hierarchy, Modeling and Advanced Planning Systems. In: de Kok AG, Graves SC (eds) Supply Chain Management: Design, Coordination and Operation. Elsevier, Amsterdam et al., pp 457-523

Fleischmann B, Meyr H, Wagner M (2005) Advanced Planning. In: Stadtler H, Kilger C (eds) Supply Chain Management and Advanced Planning, 3rd edn. Springer, Berlin et al., pp 81-106

Fong CO, Srinivasan V (1986) The Multiregion Dynamic Capacity Expansion Problem: An Improved Heuristic. Management Science 32: 1140-1152

Fong CO, Srinivasan V (1981) The Multiregion Dynamic Capacity Expansion Problem, Part I. Operations Research 29: 787-799

Forman EH (1990) Multi criteria decision making and the analytic hierarchy process. In: Bana e Costa CA (ed) Readings in Multiple Criteria Decision Aid. Springer, Berlin et al., pp 295-318

Forman EH, Gass SI (2001) The analytic hierarchy process - an exposition. Operations Research 49: 469-486

Forman E, Selly MA (2002) Decision By Objectives. World Scientific, Singapore et al.

Fraedrich D (2001) Zoll-Leitfaden für die Betriebspraxis, 12th edn. Erich Schmidt, Berlin

Francis RL, McGinnis LF, White JA (1983) Locational Analysis. European Journal of Operational Research 12: 220-252

Francis RL, McGinnis LF, White JA (1992) Facility layout and location: an analytical approach. Prentice-Hall, Englewood Cliffs

Freidenfelds (1981) Capacity Expansion. North Holland, New York and Oxford

Gallagher JR (1967) Rapid Estimation of Plant Costs. Chemical Engineering 74: 89-97

Ganeshan R, Jack E, Magazine MJ, Stephens P (1999) A taxonomic review of supply chain management research. In: Tayur S, Ganeshan R, Magazine M. (eds) Quantitative models for supply chain management. Kluwer Academic Publishers, Boston et al., pp 839-879

Geldermann J, Zhang K (2001) Software Review: "Decision Lab 2000". Journal of Multicriteria Decision Analysis 10: 317-323

Geoffrion AM (1978) Better distribution planning with computer models. In: Hax AC (ed) Studies in Operations Management. North-Holland Publishing Company, Amsterdam et al., pp 163-178

Geoffrion AM, Graves GW (1974) Multicommodity distribution system design by benders decomposition. Management Science 20: 822-844

Geoffrion AM, Graves GW, Lee S-J (1978) Strategic distribution system planning: a status report. In: Hax AC (ed) Studies in Operations Management. North-Holland Publishing Company, Amsterdam et al., pp 179-204

Geoffrion AM, Powers RF (1995) Twenty Years of Strategic Distribution System Design: An Evolutionary Perspective. Interfaces 25: 105-127

Geoffrion AM, Powers RF (1980) Facility location analysis is just the beginning (if you do it right). Interfaces 10: 22-30

Geschka H (1999) Die Szenariotechnik in der strategischen Unternehmensplanung. In: Hahn D, Taylor B (eds) Strategische Unternehmungsplanung - Strategische Unternehmungsführung, 8th edn. Physica-Verlag, Heidelberg, pp 518-545

Geunes J, Pardalos PM (2003) Network Optimization in Supply Chain Management and Financial Engineering: An Annotated Bibliography. Networks 42: 66-84

Ghosh A, Harche F (1993) Location-allocation models in the private sector: progress, problems, and prospects. Location Science 1: 81-106

Godet M, Roubelat F (1996) Creating the Future: The Use and Misuse of Scenarios. Long Range Planning 29: 164-171

Goetschalckx M (2000) Strategic Network Planning. In: Stadtler H, Kilger C (eds) Supply Chain Management and Advanced Planning. Springer, Berlin et al., pp 79-95

Goetschalckx M, Fleischmann B (2005) Strategic Network Planning. In: Stadtler H, Kilger C (eds) Supply Chain Management and Advanced Planning, 3rd edn. Springer, Berlin et al., pp 117-137

Goetschalckx M, Vidal CJ, Dogan K (2002) Modeling and design of global logistics systems: A review of integrated strategic and tactical models and design algorithms. European Journal of Operational Research 143: 1-18

Goette T (1994) Standortpolitik internationaler Unternehmen. Deutscher Universitäts-Verlag, Wiesbaden

Golany B, Roll Y (1989) An Application Procedure for DEA. Omega International Journal of Management Science 17: 237-250

Golden BL, Wasil EA, Levy DE (1989) Applications of the Analytic Hierarchy Process: A categorized, annotated bibliography. In: Golden BL, Wasil EA, Harker PT (eds) The Analytic Hierarchy Process. Springer, Berlin et al., pp 37-58

Goodwin P, Wright G (2004) Decision Analysis for Management Judgement, 3rd edn. John Wiley & Sons, Chichester et al.

Götz U, Mikus B (2002) Standorte als Ressourcen - Implikationen des "Resource Based View" für die Standortplanungslehre. Zeitschrift für Planung 13: 401-429

Götze U (2002) Standortplanung. In: Küpper H-U, Wagenhofer A (eds) Handwörterbuch Unternehmensrechnung und Controlling. Schäfer-Poeschel, Stuttgart, column 1819-1828

Götze U (1995) Standortstrukturgestaltung internationaler Industrieunternehmen. Unpublished habilitation, University of Göttingen

Götze U, Bloech J (2002) Investitionsrechnung, 3rd edn. Springer, Berlin et al.

Goumas M, Lygerou V (2000) An extension of the PROMETHEE method for decision making in fuzzy environment: Ranking of alternative energy exploitation projects. European Journal of Operational Research 123: 606-613

Grabow B, Henckel D, Hollbach-Grömig B (1995) Weiche Standortfaktoren. Kohlhammer, Stuttgart et al.

Graves SC, Willems SP (2003) Supply Chain Design: Safety Stock Placement and Supply Chain Configuration. In: de Kok AG, Graves S C (eds) Supply Chain Management: Design, Coordination and Operation. Elsevier, Amsterdam et al., pp 95-132

Grunow M, Günther H-O, Burdenik H (2006) Advanced Planning for the Design of a Global Production Network for Electrical Components. In: Dessouky M, Trappey A, Tzeng GH, Sheu D, Eid S (eds) Proceedings of the 36th International Conference on Computers and Industrial Engineering, ICCIE, June 20-23, 2006, Taipei, Taiwan 2568-2580

Gue KR (2003) A dynamic distribution model for combat logistics. Computers & Operations Research 30: 367-381

Günther H-O (2005) Supply Chain Management and Advanced Planning Systems: A tutorial. In: Günther H-O, Mattfeld DC, Suhl L (eds) Supply Chain Management und Logistik. Physica-Verlag, Heidelberg, pp 3-40

Günther H-O (1998) Produktion. In: Berndt R, Altobelli CF, Schuster P (eds) Springers Handbuch der Betriebswirtschaftslehre. Springer, Berlin et al., pp 317-367

Günther H-O, Tempelmeier H (2005) Produktion und Logistik, 5th edn. Springer, Berlin et al.
Günther H-O, van Beek P (2003) Advanced Planning and Scheduling Solutions in Process Industry. Springer, Berlin et al.
Gutierrez GJ, Kouvelis P (1995) A robustness approach to international sourcing. Annals of Operations Research 59: 165-193
Hack GD (1999) Site Selection for Growing Companies. Quorum Books, Westport, Connecticut and London
Haehling von Lanzenauer C, Sprung MR (1982) Developing Inflation Scenarios. Long Range Planning 15: 37-44
Hagedorn A (1994) Modellgestützte Planung und Kontrolle von Produktionsstandorten. Deutscher Universitäts-Verlag, Wiesbaden
Hahn D (1999) Strategische Unternehmungsführung - Grundkonzept. In: Hahn D, Taylor B (eds) Strategische Unternehmungsplanung - Strategische Unternehmungsführung, 8th edn. Physica-Verlag, Heidelberg, pp 38-50
Hahn D (1992) Planung, Organisation der. In: Frese E (eds) Handwörterbuch der Organisation, 3rd edn. Schäfer-Poeschel, Stuttgart, column 1978-1993
Hahn D, Hungenberg H (2001) PuK Controllingkonzepte. Gabler, Wiesbaden
Haigh RW (1990) Selecting a US Plant Location: The Management Decision Process in Foreign Companies. Columbia Journal of World Business 25: 22-31
Haigh RW (1989) Investment strategies and the plant-location decision. Praeger, New York et al.
Haines LM, (1998) Interval Judgements in the Analytic Hierarchy Process: A Statistical Perspective. In: Stewart TJ, van den Honert RC (eds) Trends in Multicriteria Decision Making. Springer, Berlin et al., pp 87-105
Hake B (1982) Der Beri-Index, ein Hilfsmittel zur Beurteilung des wirtschaftspolitischen Risikos von Auslandsinvestitionen. In: Lück W, Trommsdorff V (eds) Internationalisierung der Unternehmung als Problem der Betriebswirtschaftslehre. Erich Schmidt, Berlin, pp 463-473
Hale TS, Moberg CR (2003) Location Science Research: A Review. Annals of Operations Research 123: 21-35
Hamacher HW, Nickel S (1998) Classification of location models. Location Science 6: 229-242
Hanke JE, Wichern DW (2005) Business Forecasting, 8th edn. Prentice-Hall, Upper Saddle River

Hanne T (1998) Multikriterielle Optimierung: Eine Übersicht. Fachbereich Wirtschaftswissenschaften der FernUniversität Hagen, Hagen

Hanssmann F (1990) Quantitative Betriebswirtschaftslehre, 3rd edn. R. Oldenburg, München, Wien

Harker PT (1989) The Art and Science of Decision Making: The Analytic Hierarchy Process. In: Golden BL, Wasil EA, Harker PT (eds) The Analytic Hierarchy Process. Springer, Berlin et al., pp 3-36

Harkness J, ReVelle C (2003) Facility location with increasing production costs. European Journal of Operational Research 145: 1-13

Harrison TP (2003) Principles for the strategic design of supply chains. In: Harrison TP, Lee HL, Neale JJ (eds) The practice of supply chain management: where theory and application converge. Kluwer Academic Publishers, Boston et al., pp 3-12

Hartmann E, Naujok N, Bäumler M, Neff P, Vallerien SU (2001) Asset Management in Chemicals and Pharmaceuticals: The Value of Building Efficient Production Networks. Chemical Market Reporter 260: 20-22

Haug P (1985) A Multiple-Period, Mixed-Integer-Programming Model for Multinational Facility Location. Journal of Management 11: 83-96

Hax AC, Meal HC (1975) Hierarchical integration of production planning and scheduling. In: Geisler MA (ed) Logistics. North-Holland Publishing Company, Amsterdam et al., pp 53-69

Hayes RH, Wheelwright SC (1984) Restoring our competitive edge: competing through manufacturing. John Wiley & Sons, New York et al.

Heizer J, Render B (2005) Operations Management, 7th edn. Prentice-Hall, Upper Saddle River

Henrich PC (2002) Strategische Gestaltung von Produktionssystemen in der Automobilindustrie. Shaker Verlag, Aachen

Herzfeld E (1983) Joint Ventures. Jordan & Sons Limited, Bristol

Herzhoff M (2004) Szenario-Technik in der chemischen Industrie. Electronic publication university library of the Technical University Berlin, Berlin

Hicks DT (1999) Activity-based costing, 2nd edn. John Wiley & Sons, New York et al.

Higginson JK, Bookbinder JH (2005) Distribution centres in supply chain operations. In: Langevin A, Riopel D (eds) Logistics systems: Design and optimization. Springer, Berlin et al., pp 67-91

Hill TJ (1992) Incorporating manufacturing perspectives in corporate strategy. In: Voss CA (ed) Manufacturing Strategy. Chapman & Hall, New York et al., pp 3-11

Hindi KS, Pienkosz K (1999) Efficient solution of large scale, single-source, capacitated plant location problems. Journal of the Operational Research Society 50: 268-274

Hines Jr. JR (1999) Lessons from Behavioral Responses to International Taxation. National Tax Journal 52: 305-322

Hinojosa Y, Puerto J, Fernández FR (2000) A multiperiod two-echelon multicommodity capacitated plant location problem. European Journal of Operational Research 123: 271-291

Hinterhuber HH, Lauda K, Matzler K, Schatz D (1994) Strategische Standortplanung - Eine Fallstudie aus der Bekleidungsbranche. Journal für Betriebswirtschaft 44: 102-113

Hobbs BF (1980) A comparison of weighting methods in power plant siting. Decision Sciences 11: 725-737

Hodder JE, Dincer MC (1986) A Multifactor Model for International Plant Location and Financing under Uncertainty. Computers & Operations Research 13: 601-609

Hodder JE, Jucker JV (1985) International plant location under price and exchange rate uncertainty. Engineering Costs and Production Economics 9: 225-229

Hofer CW, Haller TP (1980) Globescan: A way to better international risk assessment. Journal of Business Strategy 1: 41-55

Hoffman J (1997) Fine and specialty chemicals site planned for Marseilles Port area. Chemical Market Reporter 19: 5&10

Hollas JC (2004) Transfer pricing. CA Magazine 137: 49-53

Hormozi AM, Khumawala BM (1996) An improved algorithm for solving a multi-period facility location problem. IIE Transactions 28: 105-114

Huchzermeier A, Cohen MA (1996) Valuing Operational Flexibility under Exchange Rate Risk. Operations Research 44: 100-113

Huff JO, Huff AS, Thomas H (1992) Strategic Renewal and the Interaction of Cumulative Stress and Inertia. Strategic Management Journal 13: 55-75

Hummel B (1997) Internationale Standortentscheidung. Haufe, Freiburg i. Br.

Hummeltenberg W (1981) Optimierungsmethoden zur betrieblichen Standortwahl. Physica, Würzburg

Hunt JR, Koulamas CP (1989) A Model for Evaluating Potential Facility Locations on a Global Basis. S.A.M. Advanced Management Journal 54: 19-23

Hurtmanns F, Packowski J (1999) Supply-Chain-Optimierung mit SAP APO in der Chemieindustrie: Einsatzuntersuchung und Geschäftsprozeßszenarien. HMD – Praxis der Wirtschaftsinformatik 36: 58-69

Hwang C-L, Masud AYM (1979) Multiple Objective Decision Making - Methods and Applications. Springer, Berlin et al.

Hwang C-L, Yoon K (1981) Multiple Attribute Decision Making - Methods and Applications. Springer, Berlin et al.

Ihde GB (2001) Transport, Verkehr, Logistik, 3rd edn. Vahlen, München

Jackson JH (1997) The World Trading System, 2nd edn. MIT Press, Cambridge

Jacob F (2005) Quantitative Optimierung dynamischer Produktionsnetzwerke. Dissertation, Technical University Darmstadt, Darmstadt

Jayaraman V (1998) Transportation, facility location and inventory issues in distribution network design. International Journal of Operations & Production Management 18: 471-494

Jordan WC, Graves SC (1995) Principles on the Benefits of Manufacturing Process Flexibility. Management Science 41: 577-594

Jung H (2004) Allgemeine Betriebswirtschaftslehre, 9th edn. R. Oldenburg, München, Wien

Kaiser K-H (1979) Industrielle Standortfaktoren und Betriebstypenbildung. John Wiley & Sons, Berlin

Kalcsics J, Melo T, Nickel S (2000) Facility Location Decisions in Supply Chain Management. In: Inderfurth K, Schwödiauer G, Domschke W, Juhnke F, Kleinschmidt P, Wäscher G (eds) Operations Research Proceedings 1999. Springer, Berlin et al., pp 467-472

Kallrath J (2002) Combined strategic and operational planning - an MILP success story in chemical industry. OR Spectrum 24: 315-341

Kallrath J (2000) Mixed Integer Optimization in the Chemical Process Industry. Chemical Engineering Research and Design 78: 809-822

Kallrath J, Maindl TI (2006) Real Optimization with SAP APO. Springer, Berlin et al.

Karakaya F, Canel C (1998) Underlying dimensions of business location decisions. Industrial Management & Data Systems 7: 321-329

Karkazis J (1989) Facilities location in a competitive environment: A PROMETHEE based multiple criteria analysis. European Journal of Operational Research 42: 294-304

Kasa J (1995) Understanding Trends in Foreign Exchange Rates. FRBSF Weekly Letter 22-95

Kathawala Y, Gholamnezhad H (1987) New approach to facility location decisions. International Journal of Systems Science 18: 389-402

Kaufman L, Vanden Eede M (1977) A Plant and Warehouse Location Problem. Operational Research Quarterly 28: 547-554

Keeney RL (1992) Value-Focused Thinking. Harvard University Press, Cambridge, Massachusetts

Keeney RL, Raiffa H (1976) Decisions with Multiple Objectives: Preferences and Value Tradeoffs. John Wiley & Sons, New York et al.

Kelly DL, Marucheck AS (1984) Planning Horizon Results For the Dynamic Warehouse Location Problem. Journal of Operations Management 4: 279-294

Kendrick DA (1967) Programming Investment in the Process Industries. MIT Press, Cambridge

Kendrick DA, Stoutjesdijk AJ (1978) The Planning of Industrial Investment Programs. The Johns Hopkins University Press, Baltimore and London

Kilger C, Reuter B (2005) Collaborative Planning. In: Stadtler H, Kilger C, (eds) Supply Chain Management and Advanced Planning, 3rd edn. Springer, Berlin et al., pp 259-278

Kim Y-C, McElreath R (2001) Managing operating exposure: A case study of the automobile industry. Multinational Business Review 9: 21-26

Kinkel S (2004) Erfolgsfaktor Standortplanung. Springer, Berlin et al.

Klein HE, Linneman RE (1981) The Use of Scenarios in Corporate Planning - Eight Case Histories. Long Range Planning 14: 69-77

Klincewicz JG, Luss H (1987) A Dual-Based Algorithm for Multiproduct Uncapacitated Facility Location. Transportation Science 21: 198-206

Klincewicz JG, Luss H, Yu C-S (1988) A large-scale multilocation capacity planning model. European Journal of Operational Research 34: 178-190

Kline C (1976) Maximizing profits in chemicals. Chemtech 6: 110-117

Klingman D, Phillips N, Steiger D, Wirth R, Young W (1986) The Challenges and Success Factors in Implementing an Integrated Products Planning System for Citgo. Interfaces 16: 1-19

Klose A, Drexl S (2005) Facility location models for distribution system design. European Journal of Operational Research 162: 4-29

Kobrin SJ (1979) Political risk: A review and reconsideration. Journal of International Business Studies 10: 67-80

Kogut B (1985) Designing Global Strategies: Profiting from Operational Flexibility. Sloan Management Review 27: 27-38

Kogut B, Kulatilaka N (1994) Operating Flexibility, Global Manufacturing, and the Option Value of a Multinational Network. Management Science 40: 123-139

Kohn PM, Pintelon L, Van Puyvelde F (1997) Estimating Plant Construction Costs: Develop a reliable index for non-U.S. projects in the chemical process industries. Chemical Engineering 104: 98-108

Köksalan M, Süral H (1999) Efes Beverage Group Makes Location and Distribution Decisions for Its Malt Plants. Interfaces 29: 89-103

Kölbel H, Schulze J (1967) Fertigungsvorbereitung in der Chemischen Industrie. Gabler, Wiesbaden

Koller T, Goedhart M, Wessels D (2005) Valuation, 4th edn. John Wiley & Sons, Hoboken, N.J.

Kondili E, Pantelides CC, Sargent RWH (1993) A general algorithm for short-term scheduling of batch operations - I. MILP formulation. Computers & Chemcial Engineering 17: 211-227

Kouvelis P, Rosenblatt MJ (2002) A mathematical programming model for global supply chain management: Conceptual approach and managerial insights. In: Geunes J, Pardalos PM, Romeijn HE (eds) Supply Chain Management: Models, Applications, and Research Directions. Kluwer Academic Publishers, Boston et al., pp 245-277

Kouvelis P, Rosenblatt MJ, Munson CL (2004) A mathematical programming model for global plant location problems: Analysis and insights. IIE Transactions 36: 127-144

KPMG (2006) Competitive Alternatives. KPMG

Krarup J, Pruzan PM (1990) Ingredients of Locational Analysis. In: Mirchandani PB, Francis RL (eds) Discrete Location Theory. John Wiley & Sons, New York et al., pp 1-54

Krystek U, Walldorf EG (1997) Frühaufklärung länderspezifischer Chancen und Bedrohungen. In: Krystek U, Zur E (eds) Internationalisierung. Springer, Berlin et al., pp 443-463

Kuhn A, Hellingrath B (2002) Supply Chain Management. Springer, Berlin et al.

Kulkarni SS, Magazine MJ, Raturi AS (2004) Risk Pooling Advantages of Manufacturing Network Configuration. Production and Operations Management 13: 186-199

Küpper W (1982) Standortentscheidungsprozesse transnationaler industrieller Großunternehmen. In: Lück W, Trommsdorff V (eds) Internationalisierung der Unternehmung als Problem der Betriebswirtschaftslehre. Erich Schmidt, Berlin, pp 439-461

Lackes R (1983) Die Nutzwertanalyse zur Beurteilung qualitativer Investitionseigenschaften. Das Wirtschaftsstudium 11: 385-390

Lambert DM, Cooper MC (2000) Issues in Supply Chain Management. Industrial Marketing Management 29: 65-83

Lee CY (1991) An optimal algorithm for the multiproduct capacitated facility location problem with a choice of facility type. Computers & Operations Research 18: 167-182

Lee HL (2004) The Triple-A Supply Chain. Harvard Business Review 82: 102-112

Lee HL, Billington C (1995) The Evolution of Supply-Chain-Management Models and Practice at Hewlett-Packard. Interfaces 25: 42-63

Lee J-W, Swagel P (1997) Trade Barriers and Trade Flows Across Countries and Industries. The Review of Economics and Statistics 79: 372-382

Lee S-B, Luss H (1987) Multifacility-Type Capacity Expansion Planning: Algorithms and Complexities. Operations Research 35: 249-253

Leskinen P (2000) Measurement Scales and Scale Independence in the Analytic Hierarchy Process. Journal of Multi-Criteria Decision Analysis 9: 163-174

Lesney MS (2004) Paints, Pigmenty, and Dyes. Chemical & Engineering News - Enterprise of the Chemical Sciences supplement 82: 29-32

Lessard DR, Lightstone JB (1986) Volatile exchange rates can put operations at risk. Harvard Business Review 64: 107-114

Lewin AY, Seiford LM (1997) Extending the frontiers of Data Envelopment Analysis. Annals of Operations Research 73: 1-11

Li H, Clarke-Hill CM (2004) Sino-British joint ventures in China: Investment patterns and host country conditions. European Business Review 16: 44-63

Liebmann H-P (1971) Die Standortwahl als Entscheidungsproblem. Physica, Heidelberg

Lopez RA, Henderson NR (1989) The Determinants of Location Choices for Food Processing Plants. Agribusiness 5: 619-632

Lowe TJ, Wendell RE, Hu G (2002) Screening location strategies to reduce exchange rate risk. European Journal of Operational Research 136: 573-590

Lüder K (1982) Strategische Standortplanung transnationaler industrieller Großunternehmen. In: Lück W, Trommsdorff V (eds) Internationalisierung der Unternehmung als Problem der Betriebswirtschaftslehre. Erich Schmidt, Berlin, pp 415-438

Lüder K, Küpper W (1983) Unternehmerische Standortplanung und regionale Wirtschaftsförderung. Vandenhoeck & Ruprecht, Göttingen

Lüder K, Küpper W (1982) Standortplanung industrieller Großunternehmen. Hochschule für Verwaltungswissenschaften Speyer, Speyer

Lun K, Peske T (2002) Realoptionsansatz ist für die Praxis zu komplex. Frankfurter Allgemeine Zeitung, 18.03.2002, p 26

Luss H (1982) Operations Research and Capacity Expansion Problems: A Survey. Operations Research 30: 907-947

MacCarthy BL, Atthirawong W (2003) Factors affecting location decisions in international operations - a Delphi study. International Journal of Operations & Production Management 23: 794-818

MacCormack AD, Newman III LJ, Roenfield DB (1994) The New Dynamics of Global Manufacturing Site Location. Sloan Management Review 35: 69-80

Maritan CA, Brush TH, Karnani AG (2004) Plant roles and decision autonomy in multinational networks. Journal of Operations Management 22: 489-503

Markovitz HM (1959) Portfolio Selection: Efficient diversification of investments. John Wiley & Sons, New York et al.

Martel A (2005) The design of production-distribution networks: a mathematical programming approach. In: Geunes J, Pardalos PM (eds) Supply Chain Optimization. Springer, Berlin et al., pp 265-305

Martel A (2005) The design of production-distribution networks: A mathematical programming approach. In: Geunes J, Pardalos PM (eds) Supply Chain Optimization. Springer, Berlin et al., pp 265-305

Martel A, M'Barek W, D'Amours S (2005) International Factors in the Design of Multinational Supply Chains: The Case of Canadian Pulp and Paper Companies. Université Laval Working Paper DT-2005-AM-3

Maßmann M (2005) Ein Ansatz zur kapazitierten stochastisch-dynamischen Facility-Location-Planung. Deutscher Universitäts-Verlag, Wiesbaden

Mazzola JB (1989) Generalized assignment with nonlinear capacity interaction. Management Science 35: 923-941

Mazzola JB, Neebe AW (1999) Lagrangian-relaxation-based solution procedures for a multiproduct capacitated facility location problem with choice of facility type. European Journal of Operational Research 115: 285-299

Mazzola JB, Schantz RH (1997) Multiple-Facility Loading Under Capacity-Based Economies of Scope. Naval Research Logistics 44: 229-256

McDonald B (1997) The World Trading System. St. Martin's Press, Inc., New York

McGrath ME, Hoole RW (1992) Manufacturing's New Economies of Scale. Harvard Business Review 70: 95-102

McMillan BG, Humphreys KK (1990) Sources of International Cost Data. Transactions of the American Association of Cost Engineers, pp K.6.1-K.6.9

McPadden M (2001) Mitsubishi Plans to Invest $180 Million For Polycarbonate, Polyacetal Expansions. Chemical Market Reporter 260: 18

Melachrinoudis E, Cullinane TP (1986) Locating an undesirable facility with a minimax criterion. European Journal of Operational Research 24: 239-246

Melachrinoudis E, Min H (2000) The dynamic relocation and phase-out of a hybrid, two-echelon plant/warehouse facility: A multiple objective approach. European Journal of Operational Research 123: 1-15

Meldrum DH (2000) Country Risk and Foreign Direct Investment. Business Economics 35: 33-40

Melo MT, Nickel S, Saldanha da Gama F (2005) Dynamic multi-commodity capacitated facility location: a mathematical modeling framework for strategic supply chain planning. Computers & Operations Research 33: 181-208

Meyer T (2006) Globale Standortwahl - Einflußfaktoren. In: Abele E, Kluge J, Näher U (eds) Handbuch Globale Produktion. Carl Hanser, München und Wien, pp 36-100

Meyer T (2005) Globale Produktionsnetzwerke - Ein Ansatz zur Entwicklung strategischer Standortkonzepte. Dissertation, Technical University Darmstadt, Darmstadt

Meyer T, Jacob F (2006) Gestaltung globaler Produktionsnetzwerke. In: Abele E, Kluge J, Näher U (eds) Handbuch Globale Produktion. Carl Hanser, München und Wien, pp 144-198

Meyer-Lindemann HU (1951) Typologie der Theorien des Industriestandorts. Walter Dorn, Bremen-Horn

Meyr H, Wagner M, Rohde J (2005) Structure of Advanced Planning Systems. In: Stadtler H, Kilger C (eds) Supply Chain Management and Advanced Planning, 3rd edn. Springer, Berlin et al., pp 109-115

Micallef JV (1981) Political Risk Assessment. Columbia Journal of World Business 16: 47-61

Miller T (2002) Hierarchical Operations and Supply Chain Planning. Springer, Berlin et al.

Millet I, Harker PT (1990) Globally effective questioning in the Analytic Hierarchy Process. European Journal of Operational Research 48: 88-97

Min H, Melachrinoudis E (1999) The relocation of a hybrid manufacturing/distribution facility from supply chain perspectives: a case study. Omega International Journal of Management Science 27: 75-85

Minor MS (1994) The demise of expropriation as an instrument of LDC Policy, 1980-1992. Journal of International Business Studies 25: 177-188

Mintzberg H, Raisinhani D, Théorêt A (1976) The Structure of "Unstructured" Decision Processes. Administrative Science Quarterly 21: 246-275

Mirchandani PB, Francis RL (1990) Discrete Location Theory. John Wiley & Sons, New York et al.

Mo Y, Harrison TP (2005) A conceptual framework for robust supply chain design under demand uncertainty. In: Geunes J, Pardalos PM (eds) Supply Chain Optimization. Springer, Berlin et al., pp 243-263

Mobasheri F, Orren LH, Sioshansi FP (1989) Scenario Planning at Southern California Edison. Interfaces 19: 31-44

Mohamed ZM (1999) An integrated production-distribution model for a multi-national company operating under varying exchange rates. International Journal of Production Economics 58: 81-92

Moon S (1989) A Profit-maximizing Plant-loading Model with Demand Fill-rate Constraints. Journal of the Operational Research Society 40: 1019-1027

Moyer K (1996) Scenario Planning at British Airways - A Case Study. Long Range Planning 29: 172-181

Müller B (1983) Ein Verfahren zur Unterstützung der simultanen Kapazitäts- und Standortplanung für Industrieunternehmen. Zeitschrift für Betriebswirtschaft 53: 183-201

Mullin R (2004) Fine Chemicals. Chemical & Engineering News - Enterprise of the Chemical Sciences supplement 82: 41-19

Mulvey JM, Vanderbei J, Zenios SA (1995) Robust Optimization of Large-Scale Systems. Operations Research 43: 264-281

Munson CL, Rosenblatt MJ (1997) The Impact of Local Content Rules on Global Sourcing Decisions. Production and Operations Management 6: 277-290

Muriel A, Simchi-Levi D (2003) Supply Chain Design and Planning - Applications of Optimization Techniques for Strategic and Tactical Models. In: de Kok AG, Graves SC (eds) Supply Chain Management: Design, Coordination and Operation. Elsevier, Amsterdam et al., pp 17-93

Murphy JV, Goodman RW (1998) Building a Tax-Effective Supply Chain. Global Logistics & Supply Chain Strategies, available online at http://www.glscs.com/archives/11.98.taxes.htm (as of 23.09.2006)

Mustafa A, Goh M (1996) Multi-criterion Models for Higher Education Administration. Omega International Journal of Management Science 24: 167-178

NN (2005) Höchst floriert auch ohne Hoechst. Frankfurter Allgemeine Zeitung, 26.08.2005, p 14

NN (2004) Degussa to Shift Pharma Production to China. Chemical Market Reporter 265: 6; 27

NN (2003a) ChemSite. Chemical Week 165: S6-S7

NN (2003b) VII. Foreign direct investment restrictions in OECD countries. OECD Economic Outlook 76: 167-173

NN (2000) Ciba Opens New Quinacridone Plant Capping $180 Million in Investments. Chemical Market Reporter 257: 7

Neumann K, Schwindt C, Trautmann N (2002) Advanced production scheduling for batch plants in process industries. OR Spectrum 24: 251-279

Nickel U (2006) State-of-the-art Production concepts in the Chemical Industry. In: Budde F, Felcht U-H, Frankemölle H (eds) Value Creation: Strategies for the Chemical Industry, 2nd edn. Wiley-VCH, Weinheim, pp 241-256

Nickel S, Velten S, Weimerskirch G (2005) Strategische Supply-Chain Entscheidungen in der Stahlindustrie - Eine Fallstudie. In: Günther H-O, Mattfeld DC, Suhl L (eds) Supply Chain Management und Logistik. Physica-Verlag, Heidelberg, pp 157-177

Nieckels L (1976) Transfer pricing in multinational firms. John Wiley & Sons, New York et al.

Nordas HK (2004) The Global Textile and Clothing Industry post the Agreement on Textiles and Clothing. World Trade Organization Discussion Paper 5

Olbert G (1976) Der Standortentscheidungsprozeß in der industriellen Unternehmung. Dissertation, Würzburg University, Würzburg

Owen SH, Daskin MS (1998) Strategic facility location: A review. European Journal of Operational Research 111: 423-447

Ozernoy VM (1992) Choosing the "Best" Multiple Criteria Decision-Making Method. INFOR 30: 159-171

Packowski J (1996) Betriebsführungssysteme in der Chemischen Industrie. Gabler, Wiesbaden

Papageorgiou LG, Rostein GE, Shah N (2001) Strategic Supply Chain Optimization for the Pharmaceutical Industries. Industrial Engineering & Chemistry Research 40: 275-286

Paquet M, Martel A, Desaulniers G (2004) Including technology selection decisions in manufacturing network design models. International Journal of Computer Integrated Manufacturing 17: 117-125

Paraskevopoulos D, Karakitsos E, Rustem B (1991) Robust capacity planning under uncertainty. Management Science 37: 787-800

Partovi FY, Burton J, Banerjee A (1990) Application of Analytical Hierarchy Process in Operations Management. International Journal of Operations & Production Management 10: 5-19

Pavic I, Babic Z (1991) The use of the PROMETHEE method in the location choice of a production system. International Journal of Production Economics 23: 165-174

Pellerin M, Chai J-H, Yap C-M (2003) Manufacturing location decision in the pharmaceutical industry: An exploratory study from a network perspective. Proceedings of the 8th Cambridge Symposium on International Manfacturing, pp 255-278

Perridon L, Steiner M (2006) Finanzwirtschaft der Unternehmung, 13th edn. Vahlen, München

Peskin MI, Halpern D (1990) Plant Location: Non-Economic Intangibles Are Valuable. Business Forum 15: 25-29

Petkov SB, Maranas CD (1998) Design of Multiproduct Batch Plants under Demand Uncertainty with Staged Capacity Expansions. Computers & Chemcial Engineering 22: S789-S792

Phillips LD (1986) Decision analysis and its application in industry. In: Mitra G (ed) Computer Assisted Decision Making. North-Holland Publishing Company, Amsterdam et al., pp 189-197

Philpott A, Everett G (2001) Supply Chain Optimisation in the Paper Industry. Annals of Operations Research 108: 225-237

Pirkul H, Jayaraman V (1998) A multi-commodity, multi-plant, capacitated facility location problem: formulation and efficient heuristic solution. Computers & Operations Research 25: 869-878

Pirkul H, Jayaraman V (1996) Production, Transportation, and Distribution Planning in a Multi-Commodity Tri-Echelon System. Transportation Science 30: 291-302

Plastria F (2001) Static competitive facility location: An overview of optimisation approaches. European Journal of Operational Research 129: 461-471

Pöhönen M, Hämäläinen RP (1998) On the convergence of multiattribute weighting methods. Helsinki University of Technology, Systems Analysis Laboratory, working paper WP980008, pp 1-16

Pollak P (2002) Return on Capital: Plant Shutdowns - The New Wave in the Fine Chemicals Industry. Chemical Market Reporter 261: 19-21

Pomper CL (1976) International investment planning: an integrated approach. North-Holland Publishing Company, Amsterdam et al.

Pooley J (1994) Integrated Production and Distribution Facility Planning at Ault Foods. Interfaces 24: 113-121

Popp W (1983) Strategische Planung für eine multinationale Unternehmung mit gemischt-ganzzahliger Programmierung. OR Spectrum 5: 45-57

Porter ME (1990) The Competitive Advantage of Nations. The Free Press, New York

Porter ME (1986) Changing Patterns of International Competition. California Management Review 28: 9-40

Rappaport A (1998) Creating shareholder value, 2nd edn. The Free Press, New York

Rasch AA (2006) The Role of Site Services and Infrastructure for Productivity Management. In: Budde F, Felcht U-H, Frankemölle H (eds) Value Creation: Strategies for the Chemical Industry, 2nd edn. Wiley-VCH, Weinheim, pp 257-268

Remer DS, Chai LH (1993a) Process plants, Costs of Scaled-up Units. In: McKetta JJ, Cunningham WA (eds) Encyclopedia of Chemical Processing and Design. Marcel Dekker, Inc., New York et al., pp 14-39

Remer DS, Chai LH (1993) Process Equipment, Cost Scale-up. In: McKetta JJ, Cunningham WA (eds) Encyclopedia of Chemical Processing and Design. Marcel Dekker, Inc., New York et al., pp 306-317

Remer DS, Chai LH (1990) Design Cost Factors for Scaling-up Engineering Equipment. Chemical Engineering Progress 86: 77-82

Remer DS, Idrovo J (1993) Process Equipment Cost, Biotechnology and Pharmaceutical. In: McKetta JJ, Cunningham WA (eds) Encyclopedia of Chemical Processing and Design. Marcel Dekker, Inc., New York et al., pp 294-306

Remer DS, Mattos FB (2003) Cost and scale-up factors, international inflation indexes and location factors. International Journal of Production Economics 84: 1-16

ReVelle C (2001) Location Analysis. In: Gass SI, Harris CM (eds) Encyclopedia of Operations Reserach and Management Science, 2nd edn. Kluwer Academic Publishers, Boston et al., pp 459-465

ReVelle CS, Eiselt HA (2005) Location analysis: A synthesis and survey. European Journal of Operational Research 165: 1-19

Richbell SM, Watts HD (2000) Plant closures in multiplant manufacturing firms: adding an international perspective. Management Decision 38: 80-88

Riggert T (1992) Umstrukturierung jetzt oder nie. Chemische Industrie 44: 9-10

Robock SH (1971) Political risk: identification and assessment. Columbia Journal of World Business 6: 6-20

Rohm and Haas (2006) Annual report 2005.

Roodman GM, Schwarz LB (1975) Optimal and Heuristic Facility Phase-out Strategies. AIIE Transactions 7: 177-184

Root FR (1972) Analyzing Political Risks in International Business. In: Kapoor A, Grub PD (eds) The Multinational Enterprise in Transition. Darwin Press, Princeton, pp 354-365

Rosenfield DB (1996) Global and variable cost manufacturing systems. European Journal of Operational Research 95: 325-343

Rosenhead J, Elton M, Gupta SK (1972) Robustness and Optimality as Criteria for Strategic Decisions. Operational Research Quarterly 23: 413-431

Roth M (2005) Das Vorbild für Nanjing ist Ludwigshafen. Frankfurter Allgemeine Zeitung, 29.06.2005, p 20

Roy B (2005) Paradigms and challenges. In: Figueira J, Greco S, Ehrgott M (eds) Multiple Criteria Decision Analysis: State of the Art Surveys. Springer, Berlin et al., pp 3-24

Roy B (1999) Decision-Aiding Today: What Should we Expect. In: Gal T, Stewart TJ, Hanne T (eds) Multicriteria Decision Making: Advances in MCDM Models, Algorithms, Theory, and Applications. Kluwer Academic Publishers, Boston et al., pp 1.1-1.35

Roy B (1996) Multicriteria Methodology for Decision Aiding. Kluwer Academic Publishers, Boston et al.

Roy B (1993) Decision science or decision-aid science. European Journal of Operational Research 66: 184-203

Roy B (1990a) Decision-aid and decision-making. In: Bana e Costa CA, (ed) Readings in Multiple Criteria Decision Aid. Springer, Berlin et al., pp 17-35

Roy B (1990b) The outranking approach and the foundations of ELECTRE methods. In: Bana e Costa CA (ed) Readings in Multiple Criteria Decision Aid. Springer, Berlin et al., pp 155-183

Rudberg M, Olhager J (2003) Manufacturing networks and supply chains: an operations strategy perspective. Omega International Journal of Management Science 31: 29-39

Rüschenpöhler H (1958) Der Standort industrieller Unternehmungen als betriebswirtschaftliches Problem. Duncker & Humblot, Berlin

Saaty TL (2005) The analytic hierarchy and analytic network process for the measurement of intangible criteria and for decision-making. In: Figueira J, Greco S, Ehrgott M (eds) Multiple Criteria Decision Analysis: State of the Art Surveys. Springer, Berlin et al., pp 345-407

Saaty TL (1996) Decision Making with Dependence and Feedback: The Analytic Network Process. RWS Publications, Pittsburgh

Saaty TL (1994) Highlights and critical points in the theory and application of the Analytic Hierarchy Process. European Journal of Operational Research 74: 426-447

Saaty TL (1990a) Eigenvector and logarithmic least square. European Journal of Operational Research 48: 156-160

Saaty TL (1990b) How to make a decision: The Analytic Hierarchy Process. European Journal of Operational Research 48: 9-26

Saaty TL (1986) Axiomatic Foundation of the Analytic Hierarchy Process. Management Science 32: 841-855

Saaty TL (1980) The Analytic Hierarchy Process. McGraw Hill, Boston et al.

Saaty TL, Vargas LG (2001) Models, Methods, Concepts & Applications of the Analytic Hierarchy Process. Kluwer Academic Publishers, Boston et al.

Sabathil T (1969) Standortprobleme internationaler Industrieunternehmungen. Dissertation, University Erlangen-Nürnberg, Nürnberg

Sabri EH, Beamon BM (2000) A multi-objective approach to simultaneous strategic and operational planning in supply chain design. Omega International Journal of Management Science 28: 581-598

Sahinidis NV (2004) Optimization under uncertainty: state-of-the-art and opportunities. Computers & Chemcial Engineering 28: 971-983

Sahinidis NV, Grossmann IE, Fornari RE, Chathrathi M (1989) Optimization Model for Long Range Planning in the Chemical Industry. Computers & Chemcial Engineering 13: 1049-1063

Sankaran JK, Raghavan NRS (1997) Locating and Sizing Plants for Bottling Propane in South India. Interfaces 27: 1-15

Sarkis J (2000) A comparative analysis of DEA as a discrete alternative multiple criteria decision tool. European Journal of Operational Research 123: 543-557

Sarkis J, Sundarraj RP (2002) Hub location at Digital Equipment Corporation: A comprehensive analysis of qualitative and quantitative factors. European Journal of Operational Research 137: 336-347

Schill CO (1990) Industrielle Standortplanung: eine theoretische Konzeption und deren praktische Anwendung. Lang, Frankfurt am Main et al.

Schmenner RW (1997) The erosion in European manufacturing. Production and Operations Management 6: 110-113

Schmenner RW (1982a) Multiplant Manufacturing Strategies Among the Fortune 500. Journal of Operations Management 2: 77-86

Schmenner RW (1982b) Making Business Location Decisions. Prentice-Hall, Englewood Cliffs

Schmenner RW (1979) Look beyond the obvious in plant location. Harvard Business Review 57: 126-132

Schmenner RW, Huber JC, Cook RL (1987) Geographic Differences and the Location of New Manufacturing Facilities. Journal of Urban Economics 21: 83-104

Schmidt G, Wilhelm WE (2000) Strategic, tactical and operational decisions in multi-national logistics networks: a review and discussion of modelling issues. International Journal of Production Research 38: 1501-1523

Schnaars SP (1987) How to Develop and Use Scenarios. Long Range Planning 20: 105-114

Schoemaker PJH (1995) Scenario Planning: A Tool for Strategic Thinking. Sloan Management Review 36: 25-40

Schoemaker PJH (1993) Multiple Scenario Development: Its Conceptual and Behavioral Foundation. Strategic Management Journal 14: 193-213

Schoemaker PJH, Waid CC (1982) An Experimental Comparison of Different Approaches to Determining Weights in Additive Utility Models. Management Science 28: 182-196

Schoner B, Wedley WC (1989) Alternative scales in AHP. In: Locket AG, Islei G (eds) Improving Decision Making in Organizations. Springer, Berlin et al., pp 345-34

Schoner B, Wedley WC, Choo EU (1993) A unified approach to AHP with linking pins. European Journal of Operational Research 64: 384-392

Seiford LM (1994) A DEA bibliography. In: Charnes A, Cooper WW, Lewin AY, Seiford LM (eds) Data Envelopment Analysis: Theory, Methodology, and Application. Kluwer Academic Publishers, Boston et al., pp 437-469

Selim HM, Askin RG, Vakharia AJ (1998) Cell formation in group technology: review, evaluation and directions for future research. Computers & Industrial Engineering 34: 3-20

Sethi AK, Sethi SP (1990) Flexibility in Manufacturing: A Survey. International Journal of Flexible Manufacturing Systems 2: 289-328

Shah N (2005) Process industry supply chains: Advances and challenges. Computers & Chemcial Engineering 29: 1225-1235

Shapiro AC (1981) Managing Political Risk: A Policy Approach. Columbia Journal of World Business 16: 63-69

Shapiro AC (1978) Capital Budgeting for the Multinational Corporation. Financial Management 7: 7-16

Shapiro JF (2001) Modeling the Supply Chain. Duxbury, Pacific Grove, CA

Shapiro JF (1999) On the connections among activity-based costing, mathematical programming models for analyzing strategic decisions, and the resource-based view of the firm. European Journal of Operational Research 118: 295-314

Shapiro JF, Singhal VM, Wagner SN (1993) Optimizing the Value Chain. Interfaces 23: 102-117

Shi Y, Gregory M (1998) International manufacturing networks - to develop global competitive capabilities. Journal of Operations Management 16: 195-214

Shulman A (1991) An Algorithm for Solving Dynamic Capacitated Plant Location Problems with Discrete Expansion Sizes. Operations Research 39: 423-436

Simchi-Levi D, Kaminsky P, Simchi-Levi E (2003) Designing and Managing the Supply Chain, 2nd edn. McGraw Hill, Boston et al.

Simchi-Levi D, Simchi-Levi E, Watson M (2003) Tactical planning for reinventing the supply chain. In: Harrison TP, Lee HL, Neale JJ (eds) The practice of supply chain management: where theory and application converge. Kluwer Academic Publishers, Boston et al., pp 13-30

Sinclair G, Klepper S, Cohen W (2000) What's Experience Got to Do With It? Sources of Cost Reduction in a Large Specialty Chemicals Producer. Management Science 46: 28-45

Singhvi SS (1987) A quantitative approach to site selection. Management Review 76: 47-52

Single LE, Kramer JL (1996) Tax policy and the location of plants and profits. Journal of Accounting Literature 15: 108-142

Sinuany-Stern Z, Friedman L (1998) DEA and the discriminant analysis of ratios for ranking units. European Journal of Operational Research 111: 470-478

Sinuany-Stern Z, Mehrez A, Hadad Y (2000) An AHP/DEA methodology for ranking decision making units. International Transactions in Operational Research 7: 109-124

Skinner W (1985) Manufacturing: The Formidable Competitive Weapon. John Wiley & Sons, New York et al.

Skinner W (1974) The focused factory. Harvard Business Review 52: 113-121

Skinner W (1969) Manufacturing - missing link in corporate strategy. Harvard Business Review 47: 136-145

Spring M, Boaden R (1997) "One more time: how do you win orders?": a critical reappraisal of the Hill manufacturing strategy framework. International Journal of Operations & Production Management 17: 757-779

Sridharan R (1995) The capacitated plant location problem. European Journal of Operational Research 87: 203-213

Srinivasulu SL (1981) Strategic Response to Foreign Exchange Risks. Columbia Journal of World Business 16: 13-23

Stadtler H (2005) Supply Chain Management - An Overview. In: Stadtler H, Kilger C (eds) Supply Chain Management and Advanced Planning, 3rd edn. Springer, Berlin et al., pp 9-35

Stafford HA (1980) Principles of Industrial Facility Location. Conway Publications, Norcross

Stalk G (1988) Time - The Next Source of Competitive Advantage. Harvard Business Review 66: 41-51

Steuer RE (1986) Multiple Criteria Optimization: Theory, Computation and Application. John Wiley & Sons, New York et al.

Stewart TJ (1996) Relationships between Data Envelopment Analysis and Multicriteria Decision Analysis. The Journal of the Operational Research Society 47: 654-665

Stewart TJ (1992) A Critical Survey on the Status of Multiple Criteria Decision Making Theory and Practice. Omega International Journal of Management Science 20: 569-586

Stillwell WG, von Winterfeld D, John RS (1987) Comparing Hierarchical and Nonhierarchical Weighting Methods for Eliciting Multiattribute Value Models. Management Science 33: 442-450

Stobaugh RB Jr (1969) Where in the world should we put that plant? Harvard Business Review 47: 129-136

Storck WJ (2004) Specialty Chemicals. Chemical & Engineering News - Enterprise of the Chemical Sciences supplement 82: 35-39

Syam SS (2000) Multiperiod Capacity Expansion in Globally Dispersed Regions. Decision Sciences 31: 173-195

Tansel BC, Francis RL, Lowe TJ (1983a) Location on Networks: A Survey. Part I: The p-Center and p-Median Problems. Management Science 29: 482-497

Tansel BC, Francis RL, Lowe TJ (1983b) Location on Networks: A Survey. Part II: Exploiting Tree Network Structure. Management Science 29: 498-511

Tcha DW, Lee BI (1984) A branch-and-bound algorithm for the multi-level uncapacitated facility location problem. European Journal of Operational Research 18: 35-43

Tempelmeier H (2006) Bestandsmanagement in Supply Chains, 2nd edn. Books on Demand, Norderstedt

Tempelmeier H (1999) Advanced Planning Systems. Industrie Management 15: 69-72

Tempelmeier H (1980) Die Bestimmung des optimalen Standorts eines Auslieferungslagers bei entfernungsabhängiger Nachfrage und nichtlinearen Transportkosten. Zeitschrift für Operations Research 24: B167-B187

Tempelmeier H (1980) Standortoptimierung in der Marketing-Logistik. Anton Hain, Königstein/Taunus

Tesch P (1980) Die Bestimmungsgründe des internationalen Handels und der Direktinvestitionen. Duncker & Humblot, Berlin

The National Academies (2006): Rising Above The Gathering Storm: Energizing and Employing America for a Brighter Economic Future. National Academies Press, Washington, D.C.

Thomas DJ, Griffin PM (1996) Coordinated supply chain management. European Journal of Operational Research 94: 1-15

Thompson RG, Singleton FD Jr, Thrall RM, Smith BA (1986) Comparative Site Evaluations for Locating a High-Energy Physics Lab in Texas. Interfaces 16: 35-49

Timmermann M (1972) Zur Standortplanung der Unternehmungen. Wirtschaftswissenschaftliches Studium 1: 388-393

Tong H-M, Walter CK (1980) An Empirical Study of Plant Location Decisions of Foreign Manufacturing Investors in the United States. Columbia Journal of World Business 15: 66-73

Tragantalerngsak S, Holt J, Rönnqvist M (2000) An exact method for the two-echelon, single-source, capacitated facility location problem. European Journal of Operational Research 123: 473-489

Triantaphyllou E (2001) Two New Cases of Rank Reversals when the AHP and Some of its Additive Variants are Used that do not Occur with the Multiplicative AHP. Journal of Multi-Criteria Decision Analysis 10: 11-25

Truong TH, Azadivar F (2005) Optimal design methodologies for configuration of supply chains. International Journal of Production Research 43: 2217-2236

Tsiakis P, Shah N, Pantelides CC (2001) Design of Multi-echelon Supply Chain Networks under Demand Uncertainty. Industrial Engineering & Chemistry Research 40: 3585-3604

Tversky A, Kahneman D (1991) Loss Aversion in Riskless Choice: A Reference-Dependent Model. The Quarterly Journal of Economics 106: 1039-1061

Tversky A, Kahneman D (1986) Rational Choice and the Framing of Decisions. The Journal of Business 59: S251-S278

UBS (2003) Prices and Earnings. UBS, Zürich

Upton DM (1995) What Really Makes Factories Flexible? Harvard Business Review 73: 74-84

Vaidya OS, Kumar S (2006) Analytic hierarchy process: An overview of applications. European Journal of Operational Research 169: 1-29

Vallerien SU, Wittemann N (2002) Return on Assets: Restructuring Global Production Networks. Chemical Market Reporter 262: 17-19

Van Roy TJ, Erlenkotter D (1982) A Dual-Based Procedure for Dynamic Facility Location. Management Science 28: 1091-1105

Vanston JJ, Frisbie WP, Lopreato SC, Poston DL (1977) Alternate Scenario Planning. Technological Forecasting and Social Change 10: 159-180

Vargas LG (1990) An overview of the Analytic Hierarchy Process and its applications. European Journal of Operational Research 48: 2-8

VCI (2006) Chemiewirtschaft in Zahlen. Verband der Chemischen Industrie e.V., Frankfurt am Main

VCI (2005) Die chemische Industrie in Deutschland. Verband der Chemischen Industrie e.V., Frankfurt am Main

VCI (2004) Weltchemiemarkt kompakt. Verband der Chemischen Industrie e.V., Frankfurt am Main

Vereecke A, Van Dierdonck R (2002) The strategic role of the plant: testing Ferdow's model. International Journal of Operations & Production Management 22: 492-514

Verter V (2002) An integrated model for facility location and technology acquisition. Computers & Operations Research 29: 583-592

Verter V, Dincer MC (1995) Facility Location and Capacity Acquisition: An Integrated Approach. Naval Research Logistics 42: 1141-1160

Verter V, Dincer MC (1995) Global Manufacturing Strategy. In: Drezner Z (ed) Facility Location. Springer, Berlin et al., pp 263-282

Verter V, Dincer MC (1992) An integrated evaluation of facility location, capacity acquisition, and technology selection for designing global manufacturing strategies. European Journal of Operational Research 60: 1-18

Vidal CJ, Goetschalckx M (2001) A global supply chain model with transfer pricing and transportation cost allocation. European Journal of Operational Research 129: 134-158

Vidal CJ, Goetschalckx M (2000) Modeling the effect of uncertainties on global logistics systems. Journal of Business Logistics 21: 95-120

Vidal CJ, Goetschalckx M (1997) Strategic production-distribution models: A critical review with emphasis on global supply chain models. European Journal of Operational Research 98: 1-18

Vidal CJ, Goetschalckx M (1996) The Role and Limitations of Quantitative Techniques in the Strategic Design of Global Logistics Systems. Georgia Institute of Technology Working Paper 96-023

Vincke P (1999) Outranking approach. In: Gal T, Stewart TJ, Hanne T (eds) Multicriteria Decision Making: Advances in MCDM Models, Algorithms, Theory, and Applications. Kluwer Academic Publishers, Boston et al., pp 11-1 - 11-29

Vincke P (1992) Multicriteria Decision-Aid. John Wiley & Sons, New York et al.

Vladimirou H, Zenios SA (1997) Stochastic linear programs with restricted recourse. European Journal of Operational Research 101: 177-192

Vokurka RJ, Flores BE (2002) Plant charter classifications and the operating homogeneity of US manufacturing plants. Industrial Management + Data Systems 102: 406-416 von Neumann J, Morgenstern O (1953) Theory of games and economic behavior, 3rd edn. Princton University Press, Princton

von Nitzsch R, Weber M (1993) The Effect of Attribute Ranges on Weights in Multiattribute Utility Measurements. Management Science 39: 937-943

von Winterfeldt D, Edwards W (1986) Decision analysis and behavioral research. Cambridge University Press, Cambridge et al.

Vos B, Akkermans H (1996) Capturing the dynamics of facility allocation. International Journal of Operations & Production Management 16: 57-70

Wagner HM (1969) Principles of Operations Research. Prentice-Hall, Englewood Cliffs

Wallace SW (1988) A two-stage stochastic facility-location problem with time-dependent supply. In: Ermoliev Y, Wets RJ-B (eds) Numerical Techniques for Stochastic Optimization. Springer, Berlin et al., pp 489-513

Wardrep BN (1985) Factors Which Play Major Roles in Location Decisions. Industrial Development 154: 8-13

Weber A (1909) Über den Standort der Industrien. JCB Mohr, Tübingen

Weber M, Borcherding K (1993) Behavioral influences on weight judgments in multiattribute decision making. European Journal of Operational Research 67: 1-12

Weber M, Eisenführ F, von Winterfeldt D (1988) The Effects of Splitting Attributes on Weights in Multiattribute Utility Measurement. Management Science 34: 431-445

Weiner B (1992) What Executives Should Know About Political Risk. Management Review 81: 19-22

Weistroffer HR, Smith CH, Narula SC (2005) Multiple criteria decision support software. In: Figueira J, Greco S, Ehrgott M (eds) Multiple Criteria Decision Analysis: State of the Art Surveys. Springer, Berlin et al., pp 981-1018

Wets RJ-B (1989) Stochastic Programming. In: Nemhauser G, Rinnooy Kan AHG, Todd MJ (eds) Optimization. North-Holland Publishing Company, Amsterdam et al., pp 573-629

Wets RJ-B (1974) Stochastic Programs with Fixed Recourse: The Equivalent Deterministic Program. SIAM Review 16: 309-339

Wilkin S (2000) Why Political Risk is Important to You. World Trade 13: 40-44

Wille C, Gabski H, Haller T, Kim H, Unverdorben L, Winter R (2004) Synthesis of pigments in a three-stage microreactor pilot plant - an experimental technical report. Chemical Engineering Journal 101: 179-185

Wouda FHE, van Beek P, van der Vorst JGAJ, Tacke H (2002) An application of mixed-integer linear programming models on the redesign of the supply network of Nutricia Dairy & Drinks Group in Hungary. OR Spectrum 24: 449-465

Wright G, Goodwin P (1999) Future-Focussed Thinking: Combining Scenario Planning with Decision Analysis. Journal of Multicriteria Decision Analysis 8: 311-321

Wu JA, Wu NL (1984) Analysing Multi-Dimensional Attributes for the Single Plant Location Problem via an Adaptation of the Analytical Hierarchy Process. International Journal of Operations & Production Management 4: 13-21

Yang G (2004) Integrierte Hierarchische Produktionsplanung in komplexen Wertschöpfungsnetzwerken der chemischen Industrie. Dissertation, Technical University Berlin, Berlin

Yang J, Lee H (1997) An AHP decision model for facility location selection. Facilities 15: 241-254

Yoon KP, Hwang C-L (1995) Multiple Attribute Decision Making. Sage Publications, Inc., Thousand Oaks et al.

Yoon K, Hwang C-L (1985) Manufacturing plant location analysis by multiple attribute decision making: part II - multi-plant strategy and plant location. International Journal of Production Research 23: 361-370

Zahedi F (1986) The Analytic Hierarchy Process - A Survey of the Method and its Applications. Interfaces 16: 96-108

Zangemeister C (1971) Nutzwertanalyse in der Systemtechnik, 2nd edn. Wittemannsche Buchhandlung, München

Zäpfel G (2000) Strategisches Produktionsmanagement. R. Oldenburg, München, Wien

Zeng AZ (2003) Global sourcing: process and design for efficient management. Supply Chain Management: An International Journal 8: 367-379

Zeng AZ (2002) An optimization framework for evaluating logistics costs in a global supply chain: an application to the commercial aviation industry. In: Geunes J, Pardalos PM, Romeijn HE (eds) Supply Chain Management: Models, Applications, and Research Directions. Kluwer Academic Publishers, Boston et al., pp 317-339

Lecture Notes in Economics and Mathematical Systems

For information about Vols. 1–505
please contact your bookseller or Springer-Verlag

Vol. 506: B. P. Kellerhals, Financial Pricing Models in Continuous Time and Kalman Filtering. XIV, 247 pages. 2001.

Vol. 507: M. Koksalan, S. Zionts, Multiple Criteria Decision Making in the New Millenium. XII, 481 pages. 2001.

Vol. 508: K. Neumann, C. Schwindt, J. Zimmermann, Project Scheduling with Time Windows and Scarce Resources. XI, 335 pages. 2002.

Vol. 509: D. Hornung, Investment, R&D, and Long-Run Growth. XVI, 194 pages. 2002.

Vol. 510: A. S. Tangian, Constructing and Applying Objective Functions. XII, 582 pages. 2002.

Vol. 511: M. Külpmann, Stock Market Overreaction and Fundamental Valuation. IX, 198 pages. 2002.

Vol. 512: W.-B. Zhang, An Economic Theory of Cities.XI, 220 pages. 2002.

Vol. 513: K. Marti, Stochastic Optimization Techniques. VIII, 364 pages. 2002.

Vol. 514: S. Wang, Y. Xia, Portfolio and Asset Pricing. XII, 200 pages. 2002.

Vol. 515: G. Heisig, Planning Stability in Material Requirements Planning System. XII, 264 pages. 2002.

Vol. 516: B. Schmid, Pricing Credit Linked Financial Instruments. X, 246 pages. 2002.

Vol. 517: H. I. Meinhardt, Cooperative Decision Making in Common Pool Situations. VIII, 205 pages. 2002.

Vol. 518: S. Napel, Bilateral Bargaining. VIII, 188 pages. 2002.

Vol. 519: A. Klose, G. Speranza, L. N. Van Wassenhove (Eds.), Quantitative Approaches to Distribution Logistics and Supply Chain Management. XIII, 421 pages. 2002.

Vol. 520: B. Glaser, Efficiency versus Sustainability in Dynamic Decision Making. IX, 252 pages. 2002.

Vol. 521: R. Cowan, N. Jonard (Eds.), Heterogenous Agents, Interactions and Economic Performance. XIV, 339 pages. 2003.

Vol. 522: C. Neff, Corporate Finance, Innovation, and Strategic Competition. IX, 218 pages. 2003.

Vol. 523: W.-B. Zhang, A Theory of Interregional Dynamics. XI, 231 pages. 2003.

Vol. 524: M. Frölich, Programme Evaluation and Treatment Choise. VIII, 191 pages. 2003.

Vol. 525: S. Spinler, Capacity Reservation for Capital- Intensive Technologies. XVI, 139 pages. 2003.

Vol. 526: C. F. Daganzo, A Theory of Supply Chains. VIII, 123 pages. 2003.

Vol. 527: C. E. Metz, Information Dissemination in Currency Crises. XI, 231 pages. 2003.

Vol. 528: R. Stolletz, Performance Analysis and Optimization of Inbound Call Centers. X, 219 pages. 2003.

Vol. 529: W. Krabs, S. W. Pickl, Analysis, Controllability and Optimization of Time-Discrete Systems and Dynamical Games. XII, 187 pages. 2003.

Vol. 530: R. Wapler, Unemployment, Market Structure and Growth. XXVII, 207 pages. 2003.

Vol. 531: M. Gallegati, A. Kirman, M. Marsili (Eds.), The Complex Dynamics of Economic Interaction. XV, 402 pages, 2004.

Vol. 532: K. Marti, Y. Ermoliev, G. Pflug (Eds.), Dynamic Stochastic Optimization. VIII, 336 pages. 2004.

Vol. 533: G. Dudek, Collaborative Planning in Supply Chains. X, 234 pages. 2004.

Vol. 534: M. Runkel, Environmental and Resource Policy for Consumer Durables. X, 197 pages. 2004.

Vol. 535: X. Gandibleux, M. Sevaux, K. Sörensen, V. T'kindt (Eds.), Metaheuristics for Multiobjective Optimisation. IX, 249 pages. 2004.

Vol. 536: R. Brüggemann, Model Reduction Methods for Vector Autoregressive Processes. X, 218 pages. 2004.

Vol. 537: A. Esser, Pricing in (In)Complete Markets. XI, 122 pages, 2004.

Vol. 538: S. Kokot, The Econometrics of Sequential Trade Models. XI, 193 pages. 2004.

Vol. 539: N. Hautsch, Modelling Irregularly Spaced Financial Data. XII, 291 pages. 2004.

Vol. 540: H. Kraft, Optimal Portfolios with Stochastic Interest Rates and Defaultable Assets. X, 173 pages. 2004.

Vol. 541: G.-y. Chen, X. Huang, X. Yang, Vector Optimization. X, 306 pages. 2005.

Vol. 542: J. Lingens, Union Wage Bargaining and Economic Growth. XIII, 199 pages. 2004.

Vol. 543: C. Benkert, Default Risk in Bond and Credit Derivatives Markets. IX, 135 pages. 2004.

Vol. 544: B. Fleischmann, A. Klose, Distribution Logistics. X, 284 pages. 2004.

Vol. 545: R. Hafner, Stochastic Implied Volatility. XI, 229 pages. 2004.

Vol. 546: D. Quadt, Lot-Sizing and Scheduling for Flexible Flow Lines. XVIII, 227 pages. 2004.

Vol. 547: M. Wildi, Signal Extraction. XI, 279 pages. 2005.

Vol. 548: D. Kuhn, Generalized Bounds for Convex Multistage Stochastic Programs. XI, 190 pages. 2005.

Vol. 549: G. N. Krieg, Kanban-Controlled Manufacturing Systems. IX, 236 pages. 2005.

Vol. 550: T. Lux, S. Reitz, E. Samanidou, Nonlinear Dynamics and Heterogeneous Interacting Agents. XIII, 327 pages. 2005.

Vol. 551: J. Leskow, M. Puchet Anyul, L. F. Punzo, New Tools of Economic Dynamics. XIX, 392 pages. 2005.

Vol. 552: C. Suerie, Time Continuity in Discrete Time Models. XVIII, 229 pages. 2005.

Vol. 553: B. Mönch, Strategic Trading in Illiquid Markets. XIII, 116 pages. 2005.

Vol. 554: R. Foellmi, Consumption Structure and Macroeconomics. IX, 152 pages. 2005.

Vol. 555: J. Wenzelburger, Learning in Economic Systems with Expectations Feedback (planned) 2005.

Vol. 556: R. Branzei, D. Dimitrov, S. Tijs, Models in Cooperative Game Theory. VIII, 135 pages. 2005.

Vol. 557: S. Barbaro, Equity and Efficiency Considerations of Public Higer Education. XII, 128 pages. 2005.

Vol. 558: M. Faliva, M. G. Zoia, Topics in Dynamic Model Analysis. X, 144 pages. 2005.

Vol. 559: M. Schulmerich, Real Options Valuation. XVI, 357 pages. 2005.

Vol. 560: A. von Schemde, Index and Stability in Bimatrix Games. X, 151 pages. 2005.

Vol. 561: H. Bobzin, Principles of Network Economics. XX, 390 pages. 2006.

Vol. 562: T. Langenberg, Standardization and Expectations. IX, 132 pages. 2006.

Vol. 563: A. Seeger (Ed.), Recent Advances in Optimization. XI, 455 pages. 2006.

Vol. 564: P. Mathieu, B. Beaufils, O. Brandouy (Eds.), Artificial Economics. XIII, 237 pages. 2005.

Vol. 565: W. Lemke, Term Structure Modeling and Estimation in a State Space Framework. IX, 224 pages. 2006.

Vol. 566: M. Genser, A Structural Framework for the Pricing of Corporate Securities. XIX, 176 pages. 2006.

Vol. 567: A. Namatame, T. Kaizouji, Y. Aruga (Eds.), The Complex Networks of Economic Interactions. XI, 343 pages. 2006.

Vol. 568: M. Caliendo, Microeconometric Evaluation of Labour Market Policies. XVII, 258 pages. 2006.

Vol. 569: L. Neubecker, Strategic Competition in Oligopolies with Fluctuating Demand. IX, 233 pages. 2006.

Vol. 570: J. Woo, The Political Economy of Fiscal Policy. X, 169 pages. 2006.

Vol. 571: T. Herwig, Market-Conform Valuation of Options. VIII, 104 pages. 2006.

Vol. 572: M. F. Jäkel, Pensionomics. XII, 316 pages. 2006

Vol. 573: J. Emami Namini, International Trade and Multinational Activity. X, 159 pages, 2006.

Vol. 574: R. Kleber, Dynamic Inventory Management in Reverse Logistics. XII, 181 pages, 2006.

Vol. 575: R. Hellermann, Capacity Options for Revenue Management. XV, 199 pages, 2006.

Vol. 576: J. Zajac, Economics Dynamics, Information and Equilibnum. X, 284 pages, 2006.

Vol. 577: K. Rudolph, Bargaining Power Effects in Financial Contracting. XVIII, 330 pages, 2006.

Vol. 578: J. Kühn, Optimal Risk-Return Trade-Offs of Commercial Banks. IX, 149 pages, 2006.

Vol. 579: D. Sondermann, Introduction to Stochastic Calculus for Finance. X, 136 pages, 2006.

Vol. 580: S. Seifert, Posted Price Offers in Internet Auction Markets. IX, 186 pages, 2006.

Vol. 581: K. Marti; Y. Ermoliev; M. Makowsk; G. Pflug (Eds.), Coping with Uncertainty. XIII, 330 pages, 2006.

Vol. 582: J. Andritzky, Sovereign Default Risks Valuation: Implications of Debt Crises and Bond Restructurings. VIII, 251 pages, 2006.

Vol. 583: I.V. Konnov, D.T. Luc, A.M. Rubinov† (Eds.), Generalized Convexity and Related Topics. IX, 469 pages, 2006.

Vol. 584: C. Bruun, Adances in Artificial Economics: The Economy as a Complex Dynamic System. XVI, 296 pages, 2006.

Vol. 585: R. Pope, J. Leitner, U. Leopold-Wildburger, The Knowledge Ahead Approach to Risk. XVI, 218 pages, 2007 (planned).

Vol. 586: B.Lebreton, Strategic Closed-Loop Supply Chain Management. X, 150 pages, 2007 (planned).

Vol. 587: P. N. Baecker, Real Options and Intellectual Property: Capital Budgeting Under Imperfect Patent Protection. X, 276 pages , 2007.

Vol. 588: D. Grundel, R. Murphey, P. Panos , O. Prokopyev (Eds.), Cooperative Systems: Control and Optimization. IX, 401 pages , 2007.

Vol. 589: M. Schwind, Dynamic Pricing and Automated Resource Allocation for Information Services: Reinforcement Learning and Combinatorial Auctions. XII, 293 pages , 2007.

Vol. 590: S. H. Oda, Developments on Experimental Economics: New Approaches to Solving Real-World Problems. XVI, 262 pages, 2007.

Vol. 591: M. Lehmann-Waffenschmidt, Economic Evolution and Equilibrium: Bridging the Gap. VIII, 272 pages, 2007.

Vol. 592: A. C.-L. Chian, Complex Systems Approach to Economic Dynamics. X, 95 pages, 2007.

Vol. 593: J. Rubart, The Employment Effects of Technological Change: Heterogenous Labor, Wage Inequality and Unemployment. XII, 209 pages, 2007

Vol. 594: R. Hübner, Strategic Supply Chain Management in Process Industries: An Application to Specialty Chemicals Production Network Design. XII, 243 pages, 2007

Vol. 595: H. Gimpel, Preferences in Negotiations: The Attachment Effect. XIV, 268 pages, 2007

Vol. 596: M. Müller-Bungart, Revenue Management with Flexible Products: Models and Methods for the Broadcasting Industry.XXI, 297 pages, 2007

SPRINGER NATURE

GPSR Compliance

The European Union's (EU) General Product Safety Regulation (GPSR) is a set of rules that requires consumer products to be safe and our obligations to ensure this.

If you have any concerns about our products, you can contact us on ProductSafety@springernature.com

In case Publisher is established outside the EU, the EU authorized representative is:

Springer Nature Customer Service Center GmbH
Europaplatz 3
69115 Heidelberg, Germany

FSC
www.fsc.org
MIX
Papier | Fördert
gute Waldnutzung
FSC® C083411

Kolibri 360 GmbH
Ferdinand-Jühlke-Straße 7
99095 Erfurt, Deutschland
produktsicherheit@kolibri360.de